S0-CQS-908

PROCEEDINGS OF SYMPOSIA IN APPLIED MATHEMATICS

VOLUME 1 NON-LINEAR PROBLEMS IN MECHANICS OF CONTINUA
Edited by E. Reissner (Brown University, August 1947)

VOLUME 2 ELECTROMAGNETIC THEORY
Edited by A. H. Taub (Massachusetts Institute of Technology, July 1948)

VOLUME 3 ELASTICITY
Edited by R. V. Churchill (University of Michigan, June 1949)

VOLUME 4 FLUID DYNAMICS
Edited by M. H. Martin (University of Maryland, June 1951)

VOLUME 5 WAVE MOTION AND VIBRATION THEORY
Edited by A. E. Heins (Carnegie Institute of Technology, June 1952)

VOLUME 6 NUMERICAL ANALYSIS
Edited by J. H. Curtiss (Santa Monica City College, August 1953)

VOLUME 7 APPLIED PROBABILITY
Edited by L. A. MacColl (Polytechnic Institute of Brooklyn, April 1955)

VOLUME 8 CALCULUS OF VARIATIONS AND ITS APPLICATIONS
Edited by L. M. Graves (University of Chicago, April 1956)

VOLUME 9 ORBIT THEORY
Edited by G. Birkhoff and R. E. Langer (New York University, April 1957)

VOLUME 10 COMBINATORIAL ANALYSIS
Edited by R. Bellman and M. Hall, Jr. (Columbia University, April 1958)

VOLUME 11 NUCLEAR REACTOR THEORY
Edited by G. Birkhoff and E. P. Wigner (New York City, April 1959)

VOLUME 12 STRUCTURE OF LANGUAGE AND ITS MATHEMATICAL ASPECTS
Edited by R. Jakobson (New York City, April 1960)

VOLUME 13 HYDRODYNAMIC INSTABILITY
Edited by R. Bellman, G. Birkhoff, C. C. Lin (New York City, April 1960)

VOLUME 14 MATHEMATICAL PROBLEMS IN THE BIOLOGICAL SCIENCES
Edited by R. Bellman (New York City, April 1961)

VOLUME 15 EXPERIMENTAL ARITHMETIC, HIGH SPEED COMPUTING, AND MATHEMATICS
Edited by N. C. Metropolis, A. H. Taub, J. Todd, C. B. Tompkins (Atlantic City and Chicago, April 1962)

VOLUME 16 STOCHASTIC PROCESSES IN MATHEMATICAL PHYSICS AND ENGINEERING
Edited by R. Bellman (New York City, April 1963)

OPERATIONS RESEARCH
Mathematics and Models

PROCEEDINGS OF SYMPOSIA
IN APPLIED MATHEMATICS
Volume 25

OPERATIONS RESEARCH
Mathematics and Models

AMERICAN MATHEMATICAL SOCIETY
PROVIDENCE, RHODE ISLAND
1981

LECTURE NOTES PREPARED FOR THE
AMERICAN MATHEMATICAL SOCIETY SHORT COURSE

OPERATIONS RESEARCH: MATHEMATICS AND MODELS

HELD IN DULUTH, MINNESOTA
AUGUST 19–20, 1979

EDITED BY
SAUL I. GASS

The AMS Short Course Series is sponsored by the Society's Committee on Employment and Education Policy (CEEP). The Series is under the direction of the Short Course Advisory Subcommittee of CEEP.

Library of Congress Cataloging in Publication Data

American Mathematical Society Short Course,
Operations Research: Mathematics and Models (1979: Duluth, Minn.)
Operations research, mathematics and models.
(Proceedings of symposia in applied mathematics; v. 25)
"Lecture notes prepared for the American Mathematical Society Short Course, Operations Research: Mathematics and Models held in Duluth, Minnesota, August 19–20, 1979."
Includes bibliographies.
1. Operations research. I. Gass, Saul I. II. American Mathematical Society. III. Title. IV. Series.
T57.6.A47 1979 001.4'24 81-10849
ISBN 0-8218-0029-9 AACR2
ISSN 0160-7634

1980 Mathematics Subject Classification. Primary 90–01.

CONTENTS

FOREWORD

As part of its educational activities, the American Mathematical Society sponsors special topic short courses for the attendees of its annual meetings. This volume contains the revised lecture notes for the short course "Operations Research: Mathematics and Models " given on August 19-20, 1979 at the 83rd summer meeting held in Duluth, Minnesota. These lectures emphasized specific areas of operations research and the mathematics used in modeling and solving the related problems. The topics and lecturers were:

1. Mathematical Modeling of Military Conflict Situations, Seth Bonder, Vector Research, Inc.
2. Queueing Networks, Ralph L. Disney, Virginia Polytechnic Institute and State University.
3. Practical Aspects of Fishery Management Modeling, Frederick C. Johnson, National Bureau of Standards.
4. Mathematical Modeling of Health Care Delivery Systems, William P. Pierskalla, University of Pennsylvania.
5. Operations Research: Applications in Agriculture, Robert B. Rovinsky, U. S. Department of Agriculture.
6. Mathematical Modeling Applied to the Relocation of Fire Companies, Warren E. Walker, The Rand Corporation.

Each lecturer attempted to make his presentation self-contained in terms of defining the application areas and mathematics employed. The reader of the resulting notes will find that the authors, in their desire to broaden the usefulness of the published material, have, in some instances, stretched the meaning of self-contained. Thus, the reader might find that a bit of perseverance, coupled with dipping into some subsidiary references, is required to obtain the full benefits of the written discussions. However, even the casual reader will be able to ascertain how the field of operations research has contributed to the resolution of important decision problems--and how the field of applied mathematics has flourished in the guise of operations research.

Grateful appreciation is due all who helped in the running of the short course and in producing this volume: the authors, Seth Bonder, Ralph L. Disney (co-chairman of the short course), Frederick C. Johnson, William P. Pierskalla, Robert B. Rovinsky and Christine Shoemaker, and Warren E. Walker; the officers and staff of the American Mathematical Society; the members of the AMS Committee on Employment and Educational Policy; Alan J. Goldman, who encouraged us to organize the short course and helped to edit the papers; and to the usual anonymous referees whose constructive criticisms well served the authors and editor.

SAUL I. GASS
Editor
University of Maryland
National Bureau of Standards

Proceedings of Symposia in Applied Mathematics
Volume 25, 1981

MATHEMATICAL MODELING OF MILITARY CONFLICT SITUATIONS

Seth Bonder
Vector Research, Incorporated

ABSTRACT. The resolution of many decision issues (system developments, force structures, tactics and doctrine, etc.) in the Department of Defense (DOD) requires information regarding the results of potential military engagements and campaigns. This paper will describe some of the mathematical and related modeling techniques used to generate this information in the tactical (versus strategic) warfare arena. Background information regarding historical analytic structures and types of models currently employed in DOD will be presented. Specific new developments in analytic and hybrid analytic models in the past 10-15 years will be described and some numerical results of their use presented.

The birth of operations research (OR) cannot be traced to a specific day and the roots of OR are as old as science and the management function. However, its name and the first formal operations research efforts date back to the defense or military OR activities of World War II. After some 35-40 years it is my impression that the defense arena is still the most sophisticated and probably the largest user of OR kinds of analyses. The purpose of this paper is to expose you to one dimension of defense OR activities -- the modeling of tactical military conflict situations.[1] Hopefully, this will demonstrate that

[1]In general, the term "conflict" refers to antagonistic action of one population on another in order to achieve some goal or objective. This covers a wide variety of phenomena which include: epidemics, ecological systems (relations betwen organisms and the environment), prey-predator systems (attrition of fish populations), military combat between nations or tactical units, advertising and marketing competition, civil disturbances, armament/disarmament problems among nations, and social conflict (family feuds). Extensions and modifications of the mathematical structures which I shall discuss have been used in addressing problems in all of these areas.

1980 Mathematics Subject Classification 90B99

mathematical structures can be used to describe operational phenomena in addition to the classical use in describing physical science phenomena.

I will attempt to accomplish this end with somewhat of an historical developments perspective as shown in the following outline:

1.0 Introduction
 1.1 Defense Planning Issues
 1.2 Types of Tactical Warfare Models
2.0 Analytic Model Structures Prior to 1965
 2.1 Differential Structures
 2.2 Stochastic Duels
 2.3 Deficiencies of Existing Analytic Models Prior to 1965
3.0 Model Developments 1965-1975
 3.1 The Attrition Rate
 3.2 Some Theoretical Results - Homogeneous Forces
 3.3 A Battalion Level Hybrid Analytic/Simulation Engagement Model
4.0 Overview of Current Status and Development Trends

1.0 INTRODUCTION

This section of the paper contains brief descriptions of some defense planning issues in the tactical (versus strategic) warfare arena and the types of models that have been developed and used by the defense community to address them.

1.1 Defense Planning Issues

The forces and systems involved in the tactical warfare area are expensive to develop, procure, and operate. For this reason, the Department of Defense performs a large number of studies each year in order to address the broad spectrum of planning issues noted below:

- system characteristics
- system choice
- system mix
- force structure
- tactics and doctrine
- trade-off among processes
- force level

Systems characteristics is a requirements issue: to determine the capabilities that ought to be possessed by next generation systems (e.g., the range and speed of a new aircraft, detection rates for a new sensor, etc.). The system choice issue is one of selecting from comparable systems (e.g., the A-10 versus the F-4 aircraft for close air support, the Chrysler versus the

General Motors candidate for the new XM-1 tank), while the system mix issue is one of determining the numbers of noncomparable systems to perform a similar role (attack helicopter versus close air support aircraft, air defense artillery versus air interceptors for air defense). Force structure issues are concerned with the amounts and relative proportions of the different kinds of combat units (e.g., number of armored divisions versus the number of infantry divisions, number of armored divisions versus the number of tactical air wings). Tactics and doctrinal issues are concerned with choice among alternative operating procedures for employing the systems and forces. The issue of trade-offs among processes is in essence one of force structure within a unit to balance the overall capability of the unit (e.g., the trade-off between providing resources to acquire targets for fire support and providing fire support resources to attack the targets, the trade-off between resources for intelligence collection versus resources for intelligence processing). The force level issue is concerned with the question, "How much is enough?", and is, in fact, not a military planning decision but rather a legislative one involving more political/military and political dimensions.

1.2 Types of Tactical Warfare Models

Over the past twenty-five years a large number of models have been developed to assist in addressing these types of issues. With few exceptions, the models have tended to be descriptive (rather than prescriptive) in nature, describing the battle operations and providing a time history of the status of the forces throughout the battle. As a broad categorization, three types of tactical warfare models have been developed and used in the defense area: *war games*, *simulations*, and *analytic* models. Some distinctions between these models and their characteristics are noted below.

In a war game, command behavior (the decision-making process) is made by human players, whereas automated logic in the form of engagement or tactical decision rules or implicit assumptions is used in the simulation and analytic type models. The modeling (and solution) process of war games and simulations is distinctly different from that of analytic models. In the former, the tactical phenomenon of interest is decomposed into its basic activities and events, and these are organized as they would occur in reality by sequencing them in time through the use of a network structure or directed graph. Where required, descriptions of individual event outcomes or individual activities in the network are developed. Such a model (war gaming, simulation) is then solved by "acting out" the process in a step-by-step fashion through the network to generate various battle results (ground controlled, casualties, etc.).

In an analogous fashion, the development of an analytic model also begins with the decomposition and sequencing of events and activities; however, at this juncture the modeling process differs from that of war games and simulations. Rather than developing descriptions of each event and activity, analytic descriptions of event and activity *aggregates* are developed as submodels. These submodels are then integrated in a larger overall mathematical structure. In contrast to acting out the process as a solution procedure, analytic models are solved by logical mathematical or numerical operations.

In general, any type of model can be either deterministic or stochastic (probabilistic) in structure. In simplified terms, a deterministic model always produces the same output for a fixed set of inputs. In contrast, stochastic models generally require probability distributions on the inputs and produce probability distributions over the output variables. In practice, most war games are thought of as being deterministic in nature. In fact, it is highly likely that, for a fixed set of inputs, varying the players in different runs of the war game would likely produce different sets of outputs; however, I am not aware of any war games in which the same situation was replicated with different players. A large number of the simulation models used in the tactical warfare area are stochastic and are usually solved by Monte Carlo sampling methods.[1] In essence, for a fixed set of input distributions the process is acted out via Monte Carlo sampling techniques to produce a single set of outputs and then this procedure is replicated a number of times to produce *sampling* distributions on the output variables. Although it should come as no surprise to this audience, there are many in the practicing community who believe that analytic models always must be deterministic in structure. In fact, analytic models can and have been stochastic in nature: probability distributions on the inputs generate *population* probability distributions on the outputs via standard mathematical solution procedures.

Over the years these three generic model types have been used to develop models at three levels of tactical engagements: *battalion and below* (approximately 40-50 tanks, personnel carriers, attack helicopters, and other weapon systems versus two to three times this number of opposing weapons), *division/corps* level (approximately 25 battalions opposing 75-100 battalions), and *theater* level (in Europe, approximately eight NATO corps versus 20-25 Warsaw Pact corps equivalents). The battalion and smaller unit engagement models are generally used to address some of the more microscopic issues noted earlier such as system requirements, and system choice. The division and corps level models are used to address system mix, tactics and doctrine, and some of the force structure issues. The theater level models are

[1]See Wagner (1969).

used to address a number of large-scale force structure issues, logistical questions, and questions regarding allocation of theater resources to various locations and missions.

2.0 ANALYTIC MODEL STRUCTURES PRIOR TO 1965

This section of the paper reviews some of the classical deterministic and probabilistic *analytic* model structures that were developed prior to 1965 to describe military engagement situations. These were not used extensively to address defense planning issues at that time due to a number of inherent deficiencies noted at the end of this section of the lecture.

2.1 Differential Structures

Deterministic Direct-Fire Warfare

Mathematical descriptions of warfare usually attempt to predict the "state" of the system some time after the battle is initiated, i.e., the numbers and locations of surviving forces on both sides. Intuitively, the system state is a function of the initial numbers of forces, capabilities of the weapon systems, tactics employed, and characteristics of the operating environment. Most formulations assume that it is difficult to hypothesize such a mathematical function directly. Instead, in many of the deterministic approaches, it is believed that one can assume something about the *rate* at which a force is attritted in a very small time interval. This assumption regarding the rate concept leads directly to the use of differential equations as the basic mathematical structure. Although it is recognized that combat is a random process, it is commonly considered that solutions of the differential equations are "expected values."

The classic description of direct-fire warfare activity considers large numbers of units engaging on a battlefield and hypothesizes that the rate of attrition is proportional to the number of firing opposing units. Thus,

$$\frac{dn(t)}{dt} = -\alpha m(t) , \tag{1}$$

and

$$\frac{dm(t)}{dt} = -\beta n(t) , \tag{2}$$

where

$m(t), n(t)$ = the numbers of surviving Blue and Red forces at time t after the battle begins,[1]

[1]For notational convenience, the functional dependence of m and n on time throughout a battle is omitted in succeeding developments, except where its omission would be confusing to the reader.

α = the rate at which a single Blue unit defeats Red units (Red units defeated/time/Blue unit), and

β = the rate at which a single Red unit defeats Blue units.

α and β are usually referred to as attrition rates and are assumed *constant* over time, i.e., $\alpha,\beta \neq f(t)$, although they may be state dependent. They are factors that are assumed to account for all other dimensions (weapons, doctrine, environment, etc.) of combat other than force sizes and are assumed known. This formulation additionally assumes that

(1) the number of surviving units may be treated as continuous variables,
(2) there are large numbers of *homogeneous* forces,
(3) these are all exposed and within weapon range of each other,
(4) each unit knows the location of remaining units, and
(5) fire is distributed uniformly over surviving units, i.e., it is possible to recognize who is killed (perfect intelligence).

We can determine a complete description of the battle by dividing equation (1) by equation (2) and solving

$$\frac{dn}{dm} = \frac{\alpha m}{\beta n} .$$

$$\alpha(M^2 - m^2) = \beta(N^2 - n^2) \tag{3}$$

where $m(0) = M$, and $n(0) = N$. It is said that the forces fight to a draw if m and n approach zero simultaneously, or if

$$\alpha M^2 = \beta N^2 . \tag{4}$$

That is, it is considered that the condition

$$\alpha M^2 < \beta N^2 \tag{5}$$

is a prediction that the Blue force will be defeated, i.e., winning is defined as annihilating the opposing force after a long period of time.

Equations (4) and (5) suggest the advantage of concentrating one's forces in battle. That is, if the Blue system (the weapons, doctrine, environment, etc.) is four times as effective as the Red system, the latter will need only twice the initial force size for a draw.

One can easily obtain from equations (1) and (2) the time solutions

$$m = M \cosh \sqrt{\alpha\beta}\ t - \sqrt{\beta/\alpha}\ N \sinh \sqrt{\alpha\beta}\ t,$$

and

$$n = N \cosh \sqrt{\alpha\beta}\ t - \sqrt{\alpha/\beta}\ M \sinh \sqrt{\alpha\beta}\ t.$$

Figure 1a is a graphical representation of the time solution and indicates annihilation of the Blue force. This result could have been predicted from the state solution.

$\alpha M^2 = 9 < 11.6 = \beta N^2$

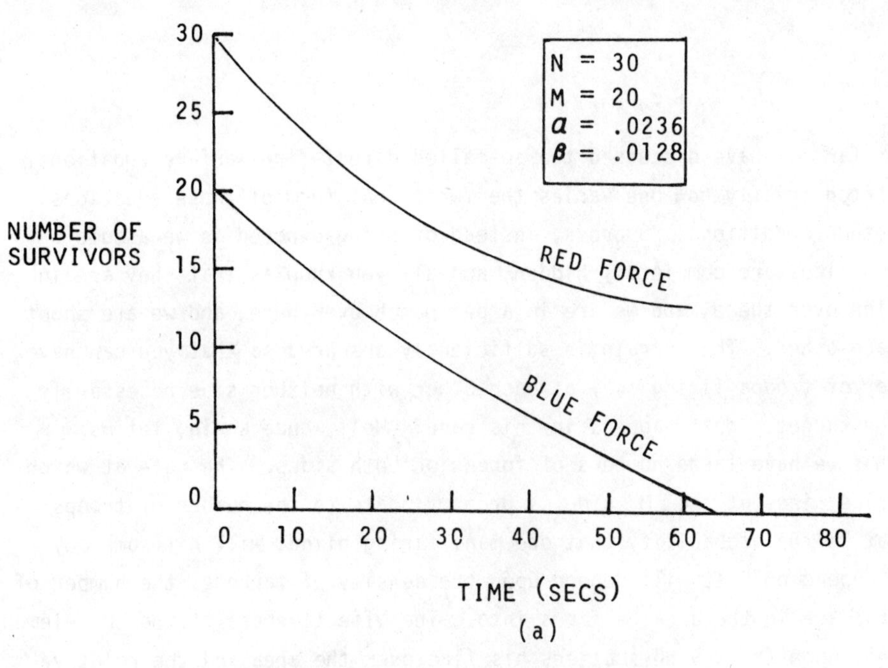

(a)

$\alpha M^2 = 20 > 11.6 = \beta N^2$

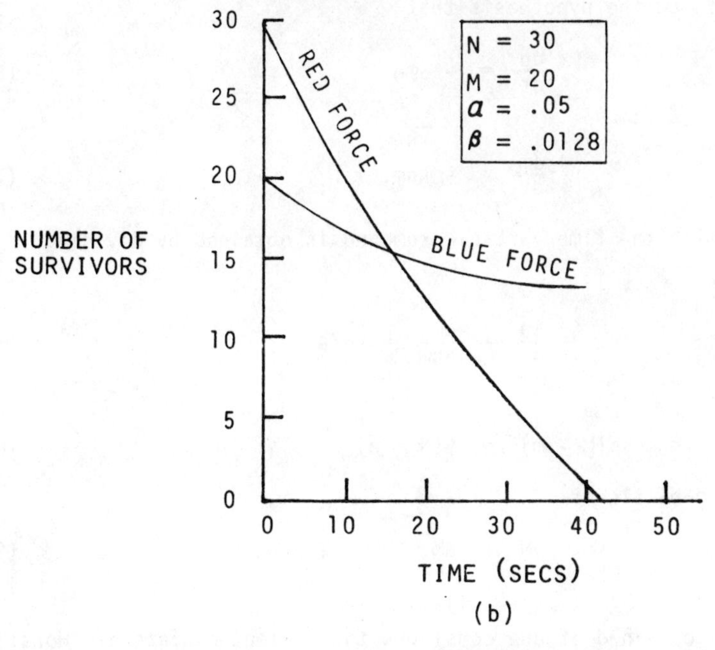

(b)

FIGURE 1

$$(.0236)\ (20)^2 = 9 < 11.6 = .0128\ (30)^2.$$

Figure 1b shows effect of increasing the Blue force attrition rate to $\alpha = .05$.

Deterministic Area Fire Warfare

Thus far, we have discussed the so-called direct-fire warfare equations. Let me sketch briefly how one varies the functional form of these equations to suit other conditions. Suppose, instead of being exposed as we assume here, both sides are completely hidden, and all you know is that they are in a pea patch over there, and we are in a pea patch over here, and we are shooting at each other. The terrain is sufficiently obscured so that you can have any number of troops firing away at each other with neither side necessarily seeing the target. What happens in this case? Well, once again, let us assume that we have large numbers of forces on both sides. The rate at which the N side fires at the M side is proportionate to the number of troops N has. What is the probability that one man, firing blind, will hit somebody going to depend on? It will depend upon the density of targets, the number of troops that are in the area he fires into. The effectiveness of the N element will depend upon the way he scatters his fire over the area and the relative amount of that area taken up by his targets, which is proportional to M, the number of targets. Assuming uniform fire over the area and uniform redistribution of targets leads to the hypothesis that

$$\frac{dn}{dt} = -\alpha mn, \tag{6}$$

and

$$\frac{dm}{dt} = -\beta nm. \tag{7}$$

The state solution with the time variable removed is obtained by dividing (6) by (7),

$$\frac{dn}{dm} = \frac{\alpha mn}{\beta nm} = \alpha/\beta$$

which leads to

$$\alpha(M - m) = \beta(N - n) \tag{8}$$

and an evenly matched battle if[1]

$$\alpha M = \beta M. \tag{9}$$

[1]This same result is obtained if one considers the ancient warfare of "Horatio at the Bridge". The forces are each in single file with only the lead men in each file engaging with swords, one against one. The numbers at any instant of time do not influence the battle.

If the Blue force effectiveness per unit is twice that of the Red, the latter will require twice the number of initial units to effect a balance of fighting strengths.

There have been many extensions and applications of these basic deterministic formulations. These include

(a) replacements (arming, when the equations are used to describe disarmament),

(b) operational losses,

(c) heterogeneous forces

$$\frac{dn_j}{dt} = - \sum_i e_{ij}\alpha_{ij}m_i$$

$$\frac{dm_i}{dt} = - \sum_j h_{ji}\beta_{ji}n_j ,$$

where $e_{ij}[h_{ji}]$ are the proportions of type $i[j]$ weapons allocated to fire on type $j[i]$ targets,

(d) guerrilla warfare,

(e) reduction in losses due to retreating, and

(f) reserve commitment.

Guerrilla Warfare

Military operations in Viet Nam (guerrilla-counterguerrilla warfare) have been modeled by a mix of the linear and square law formulations. The model considers a defender (m) force moving through an area searching for a guerrilla (n) force or intending to attack a guerrilla base. The guerrillas counter this attack by preparing an ambush for the approaching defender. When the battle starts, the guerrillas fire on the defenders who are in full view. The defenders' loss rate is thus proportional to the number of firing guerrillas. Defenders return fire blindly into the area containing the guerrillas. Accordingly, the guerrilla losses will be proportional to the number of firing regulars and the number of guerrillas occupying the area. Equations describing this situation are

$$\frac{dn}{dt} = - \alpha mn, \tag{10}$$

$$\frac{dm}{dt} = - \beta n. \tag{11}$$

It is easy to show that the state solution is

$$\beta(N - n) = \frac{\alpha}{2} (M^2 - m^2), \tag{12}$$

and the forces will be evenly matched in an engagement of this type when

$$N = \frac{\alpha}{2\beta} M^2 .$$

Assuming that this formulation is valid, early data from Viet Nam suggested that $\frac{\alpha}{2\beta} < \frac{1}{1000}$. We see from the equations that attacking guerrillas, heavily outnumbered overall, can win if both sides are subdivided into small groups and the guerrillas always attack with *local* numerical superiority. The local superiority required on the part of the guerrilla is greatly reduced if ambush tactics are used. In this case, the defenders require very large local force ratios or extremely effective weapons or both to win. The attackers (guerrillas) can counter these steps most easily by reducing their population density in their occupied area during the battle. If in a local engagement, the ambusher is approximately equal in numbers to the defender and both have weapons comparable in fighting effectiveness, then the defender probably cannot win. In practical situations, options open to both sides to change the outcome will be restricted by the limitations of resources, environment, and weaponry. Since the defender has much greater resources available than the guerrilla, the defender's use of guerrilla tactics would be a powerful tool to defeat the guerrilla force.

Probabilistic Differential Structures

The deterministic models described above miss the details of particular combats as they would develop in reality, especially for small numbers of forces. This is because combat is random in nature and develops as a stochastic process rather than a dynamic trajectory in force-size space. Although not mathematically correct, solutions of the deterministic equations are usually interpreted as expected numbers of surviving combatants, which directly imply the existence of probability distributions of the survivors.

The development of probabilistic models usually assumes that (a) m,n are treated as discrete random variables, (b) the attrition process is Markov, and (c) the process possesses a stationary transition mechanism, which may be intuitively interpreted as follows: the Markov property assumes that, from any instant of time, the behavior of the system depends on the state of the system at that instant and not on the previous history of the system. A stationary transition mechanism assumes that events (losses) which occur in a given time interval depend only on the state of the system at the beginning of the interval, and on the length of the interval -- not on the instant at which the time interval begins.

For illustrative purposes, a simplified stochastic formulation is developed below which further (but unnecessarily) assumes that in an infinitesimal time interval, Δt, (a) the probability of both forces simultaneously losing a unit is negligible, and (b) the probability of more than one loss on a side is negligible. Let

$P(m, n, t)$ = probability that there are m, n survivors after a time interval $(0,t)$,

$A\Delta t$ = probability of one Red casualty in an interval Δt,

and

$B\Delta t$ = probability of one Blue casualty in an interval Δt.

Based on the above assumptions, the system can arrive at state (m,n) after the interval $(0, t + \Delta t)$ in three mutually exclusive and collectively exhaustive ways:

(1) (m,n) survivors at time t, 0 Blue and 0 Red casualties in Δt,

(2) $(m + 1,n)$ survivors at time t, 1 Blue and 0 Red casualties in Δt, or

(3) $(m,n + 1)$ survivors at time t, 0 Blue and 1 Red casualty in Δt.

Therefore,

$$\begin{aligned} P(m,n, t + \Delta t) &= P(m,n,t)\,(1 - A\Delta t - B\Delta t) \\ &+ P(m + 1, n, t)\,(B\Delta t)\,(1 - A\Delta t) \\ &+ P(m, n + 1, t)\,(A\Delta t)\,(1 - B\Delta t). \end{aligned}$$

Expanding and recombining terms results in the relation

$$\begin{aligned} \frac{P(m,n,t + \Delta t) - P(m,n,t)}{\Delta t} &= -(A + B)\,P(m,n,t) + B\,P(m+1,n,t) \\ &+A\,P(m,n+1,t) - A\,B\Delta t\,[P(m+1,n,t) \\ &- P(m,n+1,t)]. \end{aligned}$$

Taking the limit of this equation as $\Delta t \to 0$ results in

$$\begin{aligned} \frac{d\,P(m,n,t)}{dt} &= -(A + B)\,P(m,n,t) + B\,P(m+1,n,t) \\ &+ A\,P(m,n+1,t). \end{aligned} \tag{13}$$

Equation (13), along with

$$\frac{d\,P(0, 0, t)}{dt} = A\,P(0,1,t) + B\,P(1,0,t),$$

$$\frac{d\,P(m, 0, t)}{dt} = -B\,P(m,0,t) + A\,P(m,1,t) + B\,P(m+1,0,t)$$

and

$$\frac{d\,P(0, n, t)}{dt} = -A\,P(0,n,t) + B\,P(1,n,t) + A\,P(0,n+1,t),$$

which are similarly derived, describes the time rate of change of the probabilities of different states of the system. These recursion relations can be solved subject to the initial condition that at time $t = 0$,

$$P(M,N,0) = 1,$$
$$P(m,n,t) = 0 \quad \text{for} \begin{cases} m > M \\ \text{or} \\ n > N\,. \end{cases} \tag{14}$$

Using these conditions, equation (13) reduces to

$$\frac{d\,P(M,N,t)}{dt} = -(A + B)\,P(M,N,t)$$

which, if A and B are stationary transition mechanisms (constants) gives

$$P(M,N,t) = e^{-(A+B)t}\,. \tag{15}$$

If A or B is a function of time

$$P(M,N,t) = e^{-\int_0^t (A+B)ds}\,. \tag{16}$$

We then proceed recursively to find $P(M,N-1,t)$, $P(M-1,N,t)$, $P(M-1, N-1,t)$ ... with final integration of $P(0,0,t)$.

Having (at least in principle) been able to compute $P(m,n,t)$ it is now possible to determine the probability distribution for troops on either side of the battle as a function of time and irrespective of the number of troops on the other side.

This may be done by computing

$$P_1(m,t) = \sum_{n=0}^{N} P(m,n,t), \tag{17}$$

= the probability that there are m troops at time t on the Blue side,

and

$$P_2(n,t) = \sum_{m=0}^{M} P(m,n,t), \tag{18}$$

= the probability that there are n troops at time t on the Red side.

One can then compute the average or expected numbers of troops surviving at time t by

$$\overline{m}(t) = \sum_{m=0}^{M} mP_1(m,t), \tag{19}$$

and

$$\overline{n}(t) = \sum_{n=0}^{N} nP_2(n,t). \tag{20}$$

These results may be compared with the results obtained from the corresponding set of deterministic equations (I remind you that the common interpretation of the deterministic formulations is that they predict expected numbers of troops surviving). Generally, they are not the same, and if the proper comparison is made, usually

$$\overline{m}(t) > m(t) \quad \text{and} \quad \overline{n}(t) > n(t),$$

although the differences are small for large M, N.

The probability that the Blue side wins ($m > 0$ with $n = 0$) is readily obtained if it is assumed that[1]

$$A = \frac{\alpha}{\alpha + \beta}$$

and

$$B = \frac{\beta}{\alpha + \beta} = 1 - A$$

where α and β are the constant attrition rates defined in the discussion of deterministic models. In this case, the probability that the Blue side wins may be determined by viewing the process as a sequence of Bernoulli trials (casualties) with constant probability of success A that a Red casualty occurs. The probability of at most $M - 1$ failures (Blue losses) by the Nth success (so that Red is annihilated before Blue) is given by

$$\sum_{i=0}^{M-1} \binom{N+i-1}{N-1} B^i A^{(N+i)-i} = \sum_{i=0}^{M-1} \frac{(N+i-1)!}{(N-1)!\,(i)!} B^i A^N$$

which is therefore the probability that the Blue side wins. This, of course, may be viewed as the probability of crossing the Red force = 0 barrier in a two dimensional random walk.

Finally, it is important to note that when the stationarity assumption stated by equation (15) is used, it is equivalent to the assumption of constant attrition rates employed in the deterministic formulations. That is, although the transition probabilities are at times considered state dependent, they are not considered to be functions of time as shown by equation (16) since this significantly increases the difficulty of analytic solution.

---

[1]This is analogous to the deterministic linear formulation.

2.2 Stochastic Duels

The models discussed so far have been considered macroscopic in that they consider numbers of forces and their aggregation of weapon effects in the attrition rates and transition probabilities. The theory of stochastic duels is considered microscopic because of its concern with microscopic features such as individual kill probabilities, time between rounds fired, projectile flight times, etc. This is in sharp contrast to differential models which, by omission, aggregate all of these effects.

We can obtain the flavor of such analysis by considering a simple "fundamental" duel. It is assumed that this includes

(1) Blue and Red combatants,

(2) fixed single-shot kill probabilities p_B and p_R, $(q=1-p)$,

(3) unlimited time and ammunition,

(4) projectile flight time = 0, and

(5) firing interval (T) density functions

$$f_B(t)dt = Pr(t \le T \le t + dt) = \lambda_B e^{-\lambda_B t} dt,$$

$$\frac{1}{\lambda_B} = \text{mean length of firing interval for Blue,}$$

and

$$f_R(t)dt = \lambda_R e^{-\lambda_R t} dt.$$

(6) The firing process is "uncoupled" in that the Blue combatant is firing at a passive (non-firing) Red combatant and vice versa.

By logical probability arguments, one can develop the pdf for the time for Blue to fire N rounds $T_N = t_1 + t_2 + \cdots t_N$ as

$$\frac{\lambda_B^N t^{N-1} e^{-\lambda_B t}}{(N-1)!},$$

and

$$h_B(t)dt = Pr[B \text{ kills } R \text{ in interval } (t,t+dt)|B \text{ alive}]$$

$$= p_B \left(\frac{\lambda_B t^0 e^{-\lambda_B t}}{0!} \right) dt + \ldots + q_B^{N-1} p_B \left(\frac{\lambda_B^N t^{N-1} e^{-\lambda_B t}}{(N-1)!} \right) dt + \ldots$$

$$h_B(t) = p_B \lambda_B e^{-p_B \lambda_B t}.$$

Analogously

$$h_R(t) = p_R \lambda_R e^{-p_R \lambda_R t}.$$

Then,

$$\Pr[\text{B wins at } t] = \Pr[\text{B kills R at } t|\text{alive}]P[\text{B alive at } t],$$

$$\Pr[\text{B wins}] = \int_0^\infty \left\{ h_B(t) \int_t^\infty h_R(t)dt \right\} dt,$$

$$= \frac{p_B\lambda_B}{p_B\lambda_B + p_R\lambda_R} .$$

Extensions of stochastic duel theory include

(1) limited duels (ammunition, time, both),

(2) effect of surprise allowing one opponent to fire before the other,

(3) displacement due to near miss causing missed firing turn,

(4) projectile flight time,

(5) triangular duel,

(6) square duel, and

(7) cluster duel (simultaneous firing).

2.3 Deficiencies of Existing Analytic Models Prior to 1965

Many of the analytic combat structures discussed or referred to in preceding sections were available in the middle 1960s, yet they were rarely used to address defense planning issues formulated by the advent of the Planning-Programming-Budgeting System under Secretary of Defense McNamara in 1961. The models were not used, in part, due to a number of inherent deficiencies, some of which are noted below.

Differential Structures:

- no means of predicting attrition rates (transition probabilities)
- no consideration of spatial distribution of forces
- no tactical maneuver
- constant or state dependent attrition rates only
- no consideration of terrain or environmental effects
- restrictive, matrix exponential solutions to heterogeneous force formulations

Stochastic Duels:

- difficult to model above one-on-one duel
    - omit important parameters
    - unrealistic assumptions
- no tactical maneuver
- constant or round dependent parameters only

Because of these deficiencies, and the belief that analytic type structures could not represent the complexities of land warfare, Monte Carlo simulations of small unit (e.g., battalion and below) engagements and war game models of large scale campaigns were developed and used in the sixties for operational analyses in defense planning.

3.0 MODEL DEVELOPMENTS 1965-1975

Although Monte Carlo simulation and war game models of land warfare were developed and used to assist defense planning in the sixties (and still are), it was recognized that more efficient structures would be needed to perform studies that were responsive to annual defense planning cycles. War game models at times took six to eight years to develop and, as late as 1972, one of the better division level games took six months to simulate ten hours of combat in a study. For a fixed battle situation, one of the more sophisticated battalion-level simulations required 30-60 minutes *per replication* on an IBM 370-95 computer. (Needless to say, Monte Carlo simulation approaches have not been considered to represent division, corps, or theater level battles.) Accordingly significant, albeit uncoordinated, research and development activities were conducted in the 1965-1975 time period to improve the inventory of combat models for defense planning. Some of these research and development areas are noted below.

Prediction of attrition rates (transition probabilities)
- different weapon systems
- different firing doctrines
- different types of engagements

Spatial distribution of forces
Effect of tactical maneuver
Terrain and environmental effects
Heterogeneous force formulations
- analytic
- hybrid analytic/simulation
  - independent
  - free parameter

Structured programming
Efficient solution procedures
Large scale unit (division, corps, theater) modeling
Strategy research
- optimal allocation of effort
- optimal tactical maneuver
- optimal weapon system mix

Clearly, it would be infeasible in this paper (and probably not instructive given its objective) to touch on each of the research and development efforts. Instead I will (1) describe in section 3.1 the development of one of the attrition rate models, (2) briefly display in section 3.2 some results of theoretical work on the effects of maneuver and concomitant time variations in the attrition rate, and (3) in section 3.3 describe the structure of one type of battalion-level hybrid analytic/simulation model employed heavily in defense planning. Finally, in section 4.0 I will summarize the current status of combat model developments at each of the force levels (battalion, division/corps, theater) and note some emerging trends.

3.1 The Attrition Rate

Concept of the Attrition Rate

The attrition rate for individual weapon systems was assumed to be dependent on a multitude of physical parameters of a weapon system which describe its capabilities in such areas as acquisition, firing accuracy, delivery rate, and warhead lethality. Experience with existing systems suggested that these characteristics are dependent on the range to a target and are stochastic in nature. That is, the attrition rate is functionally dependent on the range between combatants and, for any specified range, is described by a probability distribution. In the vernacular of the mathematician, the attrition rate may be viewed as a nonstationary stochastic process when forces employ mobile weapons. This is shown in figure 2, which depicts the two distinct variations in the attrition rate for a single weapon system type against one target type: (a) the stochastic variation at a specific range, which is described by the conditional probability distribution $f(\alpha|r)$, and (b) the variation in some function of the attrition-rate random variable with range, which is called the attrition-rate function, $\alpha(r)$.[1]

The fact that armed conflict is stochastic is well recognized and is one of the reasons for conceptualizing the attrition rate itself as a nonstationary stochastic process, $P[\alpha,r]$. Assuming the process $P[\alpha,r]$ could be predicted, one would like to incorporate the range and chance variations of the attrition rate explicitly into a model of combat among heterogeneous forces. The rate concept suggested that such a model would be either a differential equation (continuous-state variables) or a difference-differential equation (discrete-state variables) structure in which the relevant coefficients were nonstationary

[1]For clarity of discussion, variations in the attrition rate due to changes in target posture, environmental effect, etc., which can be included in the model, are not presented in this paper.

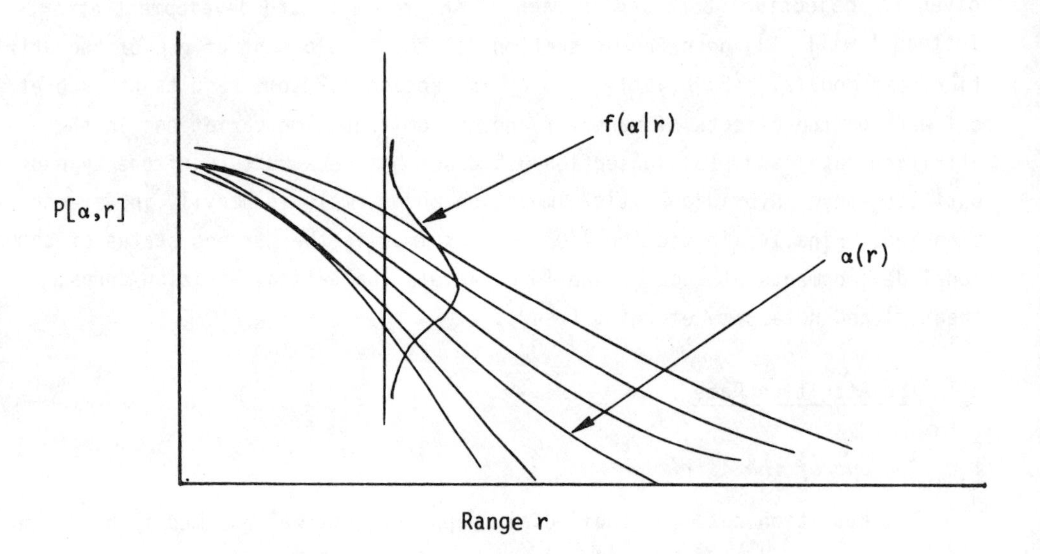

FIGURE 2 THE ATTRITION-RATE PROCESS

stochastic processes, i.e., the $P[\alpha_{ij},r]$ and $P[\beta_{ji},r]$ for all weapon-target group pairs. Initial study strongly indicated that, in the foreseeable future, there was little hope of solving either of these structures even for simplified situations. A research decision was made to suppress the chance variation in the attrition rate and to concentrate on structures of combat which explicitly involved the range variation in the rate when mobile weapons are employed.

Discrete-state stochastic process models were considered in which the transition rates are nonstationary, i.e., as varying with time. The literature indicated that discrete-state stochastic process formulations of combat have been difficult to solve even when the process is considered to be Poisson with stationary transition mechanisms. The few solutions obtained with homogeneous forces had been of such complexity as to delimit their usefulness for analysis purposes (Dolansky, 1964; Clark, 1968). Accordingly, it was felt that useful solutions for general discrete-state stochastic process formulations with non-stationary transition mechanisms could not be obtained in the near future.

Although the appropriate long-range objective is to develop stochastic formulations of heterogeneous-force armed combat such as those noted above, it was felt that a more reasonable intermediate objective would be the development of deterministic formulations, and solutions, which included the non-stationary aspects of the attrition rate at the expense of explicit considerations of its stochastic elements. Accordingly, the coupled sets of differential equations

$$\frac{dn}{dt} = -\alpha(r)m \tag{21}$$

$$\frac{dm}{dt} = -\beta(r)n \tag{22}$$

were chosen as the mathematical structure to model the combat activity *between homogeneous forces*. The nonstationary aspect of the attrition rates is included in the formulation as the variable coefficients in the differential equations, where the variable coefficients are appropriately defined as the attrition-rate functions, $\alpha(r)$ and $\beta(r)$.[1] Thus, there is one value of the attrition rate (for any firing weapon on a specific target group) at each range.

Definition of the Average Attrition Rate

The attrition rate at each range was defined to be the *harmonic mean* of the attrition-rate random variable. The appropriateness of this definition for use in a differential equation model of combat is seen below.

Consider a homogeneous-force battle in which the initial numbers of Blue (M) and Red (N) forces are sufficiently large so that neither is totally annihilated. Each Blue weapon system is engaged in a renewal process of attriting targets, i.e., the times between kills are independent and identically distributed random variables. From Blackwell's theorem (Parzen, 1962, p. 183),

$$\lim_{t\to\infty} \Pr[\text{renewal in } (t, t + dt)] = \frac{dt}{\mu},$$

where

μ = the expected interrenewal time.

Therefore, the expected number of Red kills in $(t, t + dt)$ is

$$E[\text{number of Red kills in } (t, t + dt)] = \frac{m\,dt}{\mu}. \tag{23}$$

The differential equation homogeneous-force model of combat (equation 21) states that

$$\begin{aligned} -dn &\simeq E[\text{number of Red kills in } (t, t + dt)] \\ &= \alpha m\,dt. \end{aligned} \tag{24}$$

Comparison of (23) and (24) suggests that α be defined as $1/\mu$. More generally, the definition of the attrition rate to use (for a specific range) in the differential equation structure of heterogeneous-force combat is

---

[1]Assuming a one-to-one correspondence between range r and time t.

$$\alpha_{ij}(\text{at range } r) \overset{\text{def}}{=} \frac{1}{E[T_{ij}|r]} , \tag{25}$$

where

$E[T_{ij}|r]$ = the expected time for a single Blue system of the ith group to destroy a passive j^{th} group Red target, given it is firing at a target at range r.

This definition also leads naturally to defining the range variation of the attrition rate as the variation in the reciprocal of $E[T_{ij}|r]$ as the range to the target changes. The range variation is called the *attrition-rate function* and is denoted by $\alpha_{ij}(r)$, as used in the differential equation structure of combat.

Attrition Rate for Tank Systems

In this subsection we develop an attrition-rate model for a tank gun system which is referred to as a "repeated single-shot, Markov fire" weapon system. The resultant model is the probability density function and the expected value for the *time-to-kill* random variable against a specific type target at a specific range since, by definition, it is used to predict the attrition rate. A straightforward analysis of the physical process will be used to develop the model to illustrate modeling techniques, even though more elegant mathematics such as transform techniques and Markov-renewal processes[1] can be employed. Implicit in this type of development are several assumptions which are listed here as a convenient summary and reference. These are

(a) the systems are of the impact-lethality, repeated single-shot, Markov-fire class,

(b) the probability of kill given an impact is identical for every round fired,

(c) the time preceding the firing of the first round is not random, and the conditional times to fire a second round after a hit and after a miss are not random,[2]

(d) the probability that a round fired after a preceding hit or miss results in a hit or miss is not influenced by the knowledge of other history of the engagement (such as the number of rounds fired or the number of previous hits),

(e) the engagement terminates immediately on a kill.

[1]See Bonder and Farrell (1970, Part B, Chapters 3 and 4).

[2]These assumptions can be relaxed using the transform and Markov-renewal techniques.

The firing doctrine for a main tank gun varies from round to round as shown in figure 3. Figure 3(a) shows the adjustment procedure following a hit on the first round which is to replace the crosshairs on the target--presumably the position of the crosshairs for the first round. Figure 3(b) depicts the "burst-on-target" adjustment doctrine following a miss on the first round. Succeeding adjustments, based on the result of the immediately preceding round, are made in a similar fashion until the target is defeated. The probability density function (pdf) of the time to accomplish the result is obtained by essentially modeling this adjustment process as it occurs, round by round.

Since the objective of a weapon system is to defeat the enemy, we begin by defining lethality and its unit of measurement. In brief, lethality refers to what happens to the target when struck by a projectile. The particular effect of interest is the target's combat utility. When this combat utility is reduced to zero, the target no longer poses an active tactical threat and may be considered defeated or killed. The definition of a defeated or killed target is, of course, dependent on the target's mission or role in combat. For example, consider an armored tank which is frequently referred to as "mobile, protected firepower." Some of the tank's combat missions require primarily firepower, others require mobility, and still others require both firepower and mobility, and the definition of lethality must consider which of these are relevant in the context of a study.

Lethality against a particular target is measured as the conditional probability of a kill, given the projectile hits the point target, and noted symbolically as either $P(K|H)$ or P_K. This measure is dependent on the mechanical damage caused by perforating and/or striking the target, and the loss in combat utility resulting from this mechanical damage. Procedures to predict this measure for different types of targets have been developed. See, for example, Zeller (1961), Goulet (1963), Freedman (1965), and Meyer (1967).

Another measure of lethality can be defined as "the number of hits, z, needed to defeat the target." Since we are concerned with destroying the target just once, this measure is directly related to the conditional kill probability by the geometric density function

$$p(z) = (1 - P_K)^{z-1}P_K \quad . \tag{26}$$

The number of hits needed to defeat the target, z, is initially used as a parameter in subsequent developments of this section.

The number of hits required to effect a kill describes a weapon's lethality characteristics against particular targets. The weapon's accuracy capabilities are next considered by developing the distribution for the number of rounds fired (hits and misses) to defeat the target.

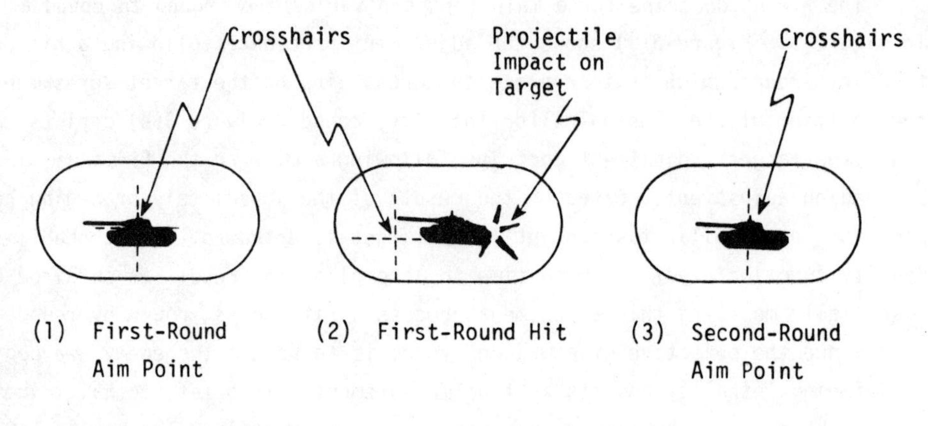

(a) Relay Following a Hit

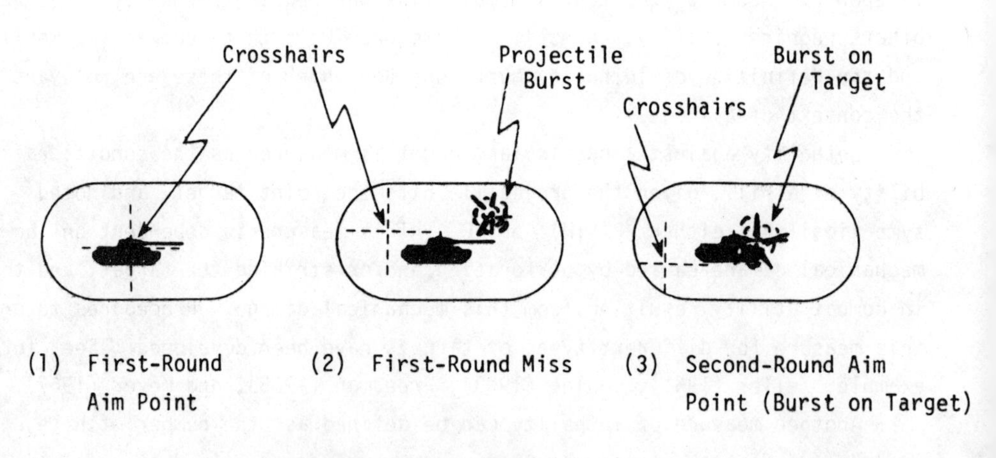

(b) Adjustment Following a Miss

FIGURE 3 FIRING DOCTRINE FOR A MAIN TANK GUN

Let

P_1 = first round hit probability,

p = conditional probability of a hit given the preceding round fired missed the target,

u = conditional probability of a hit given the preceding round fired hit the target,

and consider the sequence of trials (rounds fired) connected in a regular Markov chain with transition probability matrix

$$\begin{array}{rl} & \begin{array}{cc} \text{hit} & \text{miss} \end{array} \\ \begin{array}{r} P_1 \text{ hit} \\ (1 - P_1) \text{ miss} \end{array} & \begin{pmatrix} u & 1 - u \\ p & 1 - p \end{pmatrix} \end{array} \qquad \begin{array}{l} 0 < u < 1 \\ 0 < p < 1 \, . \end{array}$$

It is assumed that p and u are defined only on the open interval $(0,1)$. We seek the pdf for the number of rounds, N, to obtain z hits if the sequence of firings ends with a hit.[1] This can occur in two mutually exclusive and collectively exhaustive ways.

$$f(N|z) = f(N{\cdot}H{\cdot}H|z) + f(N{\cdot}M{\cdot}H|z) \; . \tag{27}$$

The first term on the right-hand side of (27) is the probability that the first and last rounds of the sequence result in hits given that the z hits occur in N firings. The second term is the probability that the first and last rounds of the sequence result in a miss and a hit, respectively, given that the z hits occur in N firings.

To determine $f(N{\cdot}H{\cdot}H|z)$ we consider the following combination of firing results:

In the first r_1 firings, the event hit occurs every time;

In the next s_1 firings, the event miss occurs every time;

In the next r_2 firings, the event hit occurs every time;

In the next s_2 firings, the event miss occurs every time;

....

In the next s_{k-1} firings, the event miss occurs every time;

In the last r_k firings, the event hit occurs every time.

---

[1]The procedure could be extended to remove this assumption that the firer recognizes when the target is defeated without technical difficulty but with increased complexity of discussion.

The joint occurrence of these events has the probability

$$P_1 u^{r_1-1}(1-u)(1-p)^{s_1-1} pu^{r_2-1}(1-u)(1-p)^{s_2-1} p \ldots pu^{r_k-1}$$

$$= P_1 u^{r_1+r_2+\ldots r_k-k}(1-u)^{k-1}(1-p)^{s_1+s_2\ldots s_{k-1}-(k-1)} p^{k-1}. \qquad (28)$$

Since there are a total of z hits and $(N - z)$ misses,

$$\sum_{i=1}^{k} r_i = z \qquad \text{and} \qquad \sum_{i=1}^{k-1} s_i = N - z .$$

Therefore, (28) becomes

$$P_1 u^{z-k}(1-u)^{k-1}(1-p)^{N-z-k+1} p^{k-1}$$

Accordingly, the probability of the outcome depends only on N, z, and k and not on the values of r_i and s_i. The number of hits, z, can be expressed as a sum of k positive integers (the r_i) in $\binom{z-1}{k-1}$ ways and the number of misses, $(N-z)$, as a sum of $(k-1)$ positive integers (the s_i) in $\binom{N-z-1}{k-2}$ ways. Therefore, the probability that it takes N firings to obtain z hits, the first and last being hits with probability P_1 and p or u, respectively--where the hits occur in k groups and the misses in $(k-1)$ groups--is

$$P_1 \binom{z-1}{k-1}\binom{N-z-1}{k-2} u^{z-k}(1-u)^{k-1} p^{k-1}(1-p)^{N-z-k+1} .$$

The outcome can occur for all values of k such that $(1 \leq k \leq z)$. Accordingly,

$$f(N\cdot H\cdot H|z) = \begin{cases} P_1 u^{z-1} & N = z \\ P_1 \displaystyle\sum_{k=2}^{z} \binom{z-1}{k-1} u^{z-k}(1-u)^{k-1} p^{k-1} \binom{N-z-1}{k-2} (1-p)^{N-z-k+1} & N > z \end{cases} \qquad (29)$$

since $\binom{N-z-1}{k-2} = 0$ when $k = 1$ and $N > z$.

By an analogous derivation, it can be shown that

$$f(N{\cdot}M{\cdot}H|z) = (1 - P_1)\sum_{k=1}^{z}\binom{z-1}{k-1} u^{z-k}(1 - u)^{k-1}p^{k}\binom{N-z-1}{k-1}(1 - p)^{N-z-k} \quad \text{for } N > z . \tag{30}$$

Substituting (29) and (30) into (27) completes the derivation for

$$f(N|z) = \begin{cases} P_1 u^{z-1} & N = z \\ \left[P_1 \sum_{k=2}^{z}\binom{z-1}{k-1} u^{z-k}(1 - u)^{k-1}p^{k-1}\binom{N-z-1}{k-2}q^{N-z-k+1} + Q_1 \sum_{k=1}^{z}\binom{z-1}{k-1} u^{z-k}(1 - u)^{k-1}p^{k}\binom{N-z-1}{k-1}q^{N-z-k}\right] & N > z , \end{cases} \tag{31}$$

where $Q_1 = (1 - P_1)$ and $q = (1 - p)$. The reader is reminded that equation (31) is a conditional distribution which is dependent on the integer z.

It is a straightforward matter to show that the characteristic function of (31) is

$$\phi_{N|z}(s) = E[e^{isN}] = \sum_{N=0}^{\infty} e^{isN} f(N|z)$$

$$= e^{isz}\left[P_1 + \frac{Q_1 pe^{is}}{1 - qe^{is}}\right]\left[u + \frac{(1 - u)pe^{is}}{1 - qe^{is}}\right]^{z-1} \tag{32}$$

where s is a dummy variable and $i = \sqrt{-1}$. Setting $s = 0$ in (32),

$$\phi_{N|z}(0) = \sum_{N=0}^{\infty} f(N|z) = 1 ,$$

proves that (31) is, in fact, a probability density function. The expected value of N is obtained from (32) as

$$E[N|z] = \frac{1}{i} \frac{d\phi_{N|z}(s)}{ds} \Bigg|_{s=0}$$

$$= z + \frac{(1 - P_1)}{p} + \frac{(1 - u)(z - 1)}{p} \quad . \tag{33}$$

The density function $f(N|z)$ for the number of rounds that must be fired to destroy a particular target is dependent on the lethality and accuracy capabilities of the weapon system. Two other important weapon characteristics remain to be considered -- the system's acquisition capabilities and its rate of fire. We consider these characteristics in a manner such that the acquisition and firing processes are serial. That is, targets are destroyed by sequentially acquiring a target, attriting it by fire, acquiring a new target, attriting it, acquiring a new target, etc. This is in contrast to parallel acquisition and firing processes in which new targets may be acquired while a previously acquired one is being attrited.

We include the timing characteristics of acquisition and firing by defining

τ_a = the time to acquire targets,

τ_1 = time to fire the first round,

τ_h = time to fire a round given the preceding round was a hit,

τ_m = time to fire a round given the preceding round was a miss,

τ_f = projectile flight time,

and consider the following sequence of events from target acquisition to destruction. The sequence begins with detection which takes τ_a time units to occur. The first round is then fired and arrives at the target area $(\tau_1 + \tau_f)$ time units later. If the first round misses, the next round will arrive $(\tau_m + \tau_f)$ time units after the first. If the first round hits the target, and more than one hit is required $(z > 1)$, the next round will arrive $(\tau_h + \tau_f)$ time units later. The sequence of firing after hits and misses is continued until the final hit which destroys the target is obtained. This description is consistent with our single-shot Markov firing doctrine in which the result of the previous round is observed before the next one is fired.

In this process, rounds will be fired after each of $(z - 1)$ hits and $(N - z)$ misses. Accordingly, the time to defeat a target may be written as

$$\begin{aligned} T &= \tau_a + (\tau_1 + \tau_f) + (\tau_h + \tau_f)(z - 1) + (\tau_m + \tau_f)(N - z) \\ &= c_1 + c_2 N , \end{aligned} \tag{34}$$

where

$$c_1 = \tau_a + \tau_1 - \tau_h + (\tau_h - \tau_m)z \tag{35}$$

$$c_2 = \tau_m + \tau_f . \tag{36}$$

Equation (34) defines T as a linear function of the discrete random variable N, and establishes a one-to-one transformation between their respective sample spaces. The density function of T is readily obtained from (31) by the change of variables technique for discrete variables as

$$f(T|z) = \begin{cases} P_1 u^{z-1} & T = c_1 + c_2 z \\ \left[P_1 \sum_{k=2}^{z} \binom{z-1}{k-1} u^{z-k}(1-u)^{k-1} p^{k-1} \binom{\left[\frac{T-c_1}{c_2}\right] - z - 1}{k-2} q^{\left(\frac{T-c_1}{c_2}\right) - z - k + 1} \right. & \\ \left. + \, Q_1 \sum_{k=1}^{z} \binom{z-1}{k-1} u^{z-k}(1-u)^{k-1} p^{k} \binom{\left[\frac{T-c_1}{c_2}\right] - z - 1}{k-1} q^{\left(\frac{T-c_1}{c_2}\right) - z - k} \right] & T > c_1 + c_2 z \end{cases} \tag{37}$$

The characteristic function of T, $\phi_{T|z}(s)$, is obtained directly from (32) and the definition of T given in (34), as

$$\begin{aligned} \phi_{T|z}(s) &= E\left[e^{isT}\right] \\ &= e^{is(c_1 + c_2 z)} \left[P_1 + \frac{Q_1 p e^{ic_2 s}}{(1 - q e^{ic_2 s})} \right] \left[u + \frac{(1 - u) p e^{ic_2 s}}{1 - q e^{ic_2 s}} \right]^{z-1} . \end{aligned} \tag{38}$$

The expected value of T can be obtained from (38), or, more directly, by employing the linear property of the expected-value operator with (34). Accordingly,

$$E[T|z] = c_1 + c_2 E[N|z]$$

$$= c_1 + c_2 \left[\frac{(1 - P_1)}{p} + \frac{(z - 1)(1 - u)}{p} + z \right] . \qquad (39)$$

The expected time to destroy a target, $E[T|z]$, is conditioned on the integer-valued lethality variable z, which is the number of hits required to destroy the target. This conditioning is removed and the continuous lethality parameter P_K (the conditional probability of destroying the target given it is hit by a projectile) introduced by

$$E[T] = \sum_{z=1}^{\infty} E[T|z]p(z)$$

$$= \tau_a + \tau_1 - \tau_h + \left(\frac{\tau_h + \tau_f}{P_K}\right) + \left(\frac{\tau_m + \tau_f}{p}\right) \left[\frac{1 - u}{P_K} + u - P_1\right] , \qquad (40)$$

where $p(z)$ is given by equation (26). Finally, the attrition rate for tank gun fire is given by

$$\alpha \overset{\text{def}}{=} \frac{1}{E[T]} \qquad (41)$$

Attrition rate models for many different types of weapon systems and firing doctrines have been developed.

3.2 Some Theoretical Results - Homogeneous Forces

Given the availability of attrition rate models as a function of basic weapon system performance capabilities (e.g., accuracy, acquisition, firing times, lethality, etc.) and information on how these capabilities change with range to a target, the attrition rates as a function of range, $\alpha(r)$ and $\beta(r)$, can be calculated directly. These are used as data in a number of models used in defense planning studies such as the one which will be described in section 3.3. Additionally, a significant amount of theoretical work has been performed with mathematical forms of $\alpha(r)$ and $\beta(r)$ in conjunction with equations (21) and (22). Thus, converting equations (21) and (22) into an appropriately defined spatial dimension and combining, produces the equation

$$\frac{d^2 n}{dr^2} + \left[\frac{w}{v^2} - \frac{1}{\alpha(r)} \frac{d\alpha(r)}{dr}\right]\frac{dn}{dr} - \left[\frac{\alpha(r)\beta(r)}{v^2}\right] n = 0 \qquad (42)$$

describing the number of surviving "Red" forces (n) as a function of the distance between forces, where

r = the range between forces
v = relative velocity between forces
w = relative acceleration between forces.

A similar equation for the surviving Blue forces (m) exists.

Solutions of equation (42) and the analogous one for the Blue forces for different attrition rate functions, $\alpha(r)$ and $\beta(r)$, have provided some insights regarding combat dynamics. For example, it is believed that the Soviets are willing to accept heavy tank losses in using velocity and mass to achieve a successful attack. Figure 4, derived from equation (42) for particular $\alpha(r)$ and $\beta(r)$, suggests that use of velocity and mass may be the *best* way to *reduce* losses if a force wishes to attack. The curves suggest that increasing the attack velocity increases the number of survivors but with decreasing marginal returns. For a fixed attack velocity, higher initial force ratios conserve attackers. In essence, speed and force concentration are good ways to saturate a defender's retaliatory capability. Some additional theoretical analyses of tactical maneuver spatial dynamics are presented in Bonder and Farrell (1970).

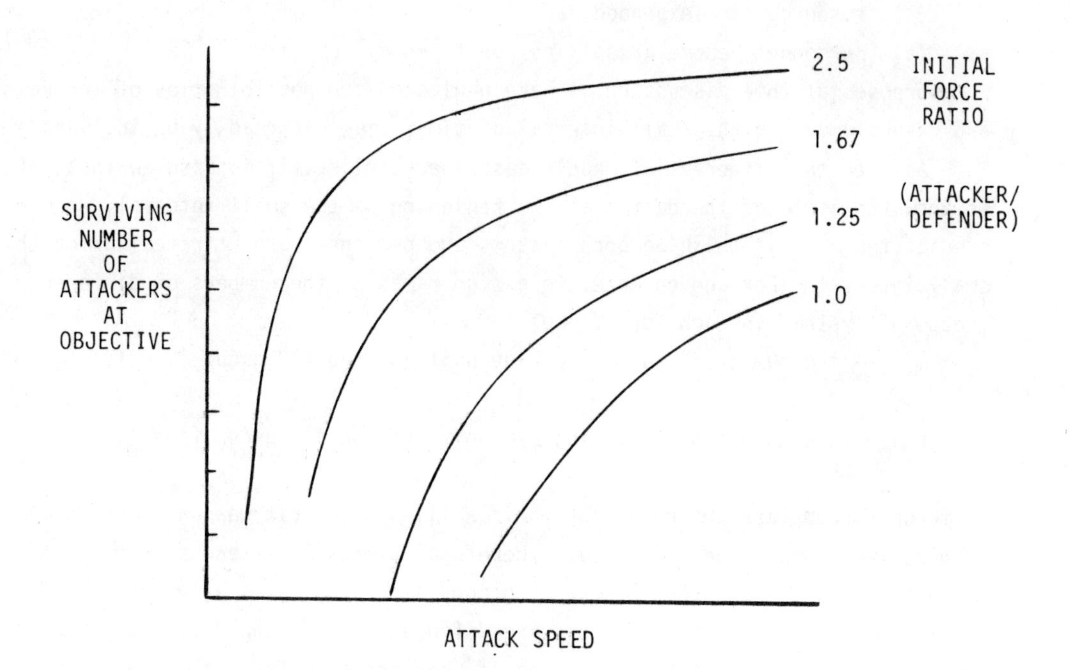

FIGURE 4 SURVIVING NUMBER OF ATTACKERS AS A FUNCTION OF ATTACK SPEED

3.3 A Battalion Level Hybrid Analytic/Simulation Engagement Model

The ability to *predict* attrition rates as a function of measurable (or predictable) weapon system capabilities has facilitated the development of a number of *hybrid analytic/simulation* models of military battles which, because of their responsiveness, have supplemented (and oftentimes replaced) Monte Carlo simulations and war games as vehicles for defense planning studies. A hybrid analytic/simulation model is a model which combines analytic and simulatory approaches to modeling the constituent processes of combat, usually employing numerical solution techniques. This section of the paper describes the structure of one of the earlier battalion level hybrid models developed for the Army in 1969. This model, and its numerous sucessors, have been used extensively to address many DOD requirements and system choice issues. In this model, the attrition, target acquisition, and intelligence processes are modeled analytically, while the movement, command and control (decision), and terrain effects processes are simulated.

The model is a deterministic differential one which focuses on small time intervals during the battle. In particular, for each side, it is hypothesized that in a short period of time

(a) locations change due to tactical movement,

(b) weapon systems are attrited by enemy activity,

(c) resources are expended, and

(d) personnel become casualties due to enemy activity.

For purposes of this discussion, I have neglected the possibilities of arrivals and resupply during the small interval of time; they can readily be included.

As with the differential models described earlier, it is assumed that, if we know the state of the battle at the beginning of the small interval, we can predict the *rate* at which weapons systems and personnel are attrited during this small interval. For convenience, we assign names to the numbers of different groups of systems in each force. Let

m_i = the number of surviving Blue units of the ith group ($i = 1, 2, \ldots, I$),

n_j = the number of surviving Red units of the jth group ($j = 1, 2, \ldots, J$).

Different groups are determined by their ability to attrit weapons systems of an opposing group or be attritted. Therefore, a missile weapon system and a rapid-fire machine gun form different groups since the rates at which they can attrit targets of an opposing group are different. Additionally, similar weapon system types can form different groups if they are at different ranges to the target and this range difference affects their ability to attrit it. Thus, a tank platoon at 1,000 meters to the target forms a different group than another tank platoon at 2,000 meters from it.

We assume that

(a) the rate of loss of units in the jth Red group due to the ith Blue group is proportional to the number of units in the ith Blue group with a proportionality factor called the *attrition coefficient*, and

(b) the rate of loss of units in the jth Red group in total is the sum of the rates of losses due to different i^{th} Blue groups.

Mathematically, these assumptions take the form of the following coupled sets of variable-coefficient differential equations to describe heterogeneous-force battles:[1]

$$\frac{dn_j}{dt} = -\sum_i A_{ij}(r_{ij})m_i \qquad \text{for } j = 1, 2, \ldots, J, \tag{43}$$

$$\frac{dm_i}{dt} = -\sum_j B_{ji}(r_{ij})n_j \qquad \text{for } i = 1, 2, \ldots, I, \tag{44}$$

where

$A_{ij}(r_{ij})$ = the utilized per system effectiveness of systems in the ith Blue group against the jth Red target group at range r. This is called the Blue attrition coefficient.

$B_{ji}(r_{ij})$ = the utilized per system effectiveness of systems in the jth Red group against the ith Blue target group at range r. This is called the Red attrition coefficient.

Although the variable r_{ij} is used to designate the range between the firing weapon group and the target group, it should be noted that, in application of the model, actual time trajectories and positions of each group are considered. Additionally, although not explicitly shown, resources expended are included in the development of the A_{ij}, and can be determined directly from the model.

It is noted that this formulation is a deterministic one which treats the numbers of surviving forces (m_i and n_j) as continuous variables, while clearly the actual battle activity is a random phenomenon and m_i and n_j are integer-valued variables. Although many probabilistic arguments are contained

[1]Battles in which at least one of the forces has more than one group.

in this formulation, the output of the model is a deterministic trajectory of the surviving numbers of forces.[1]

The attrition coefficients (A_{ij} and B_{ji}) are, as one would expect, complex functions of the weapon capabilities, target characteristics, distribution of the targets, allocation procedures for assigning weapons to targets, intelligence, etc. The model attempts to reflect these complexities by partitioning the total attrition process into four distinct ones:

(1) the effectiveness of weapons systems while firing on live targets,

(2) the allocation procedure of assigning weapons to targets,

(3) the inefficiency of fire when other than live targets are engaged, and

(4) the effect of terrain on limiting the firing activity due to loss of acquisition capability and also on mobility of the systems.

These effects are included in the attrition coefficient as

$$A_{ij}(r_{ij}) = \alpha_{ij}(r_{ij})e_{ij}(r_{ij})I_{ij}(r_{ij})F_{ij}(t) \tag{45}$$

$$B_{ji}(r_{ij}) = \beta_{ji}(r_{ij})h_{ji}(r_{ij})K_{jk}(r_{ij})G_{ji}(t) \tag{46}$$

where

$\alpha_{ij}(r)$ = *the attrition rate.* The rate at which an individual system in the ith Blue group destroys live j^{th}-group Red targets at range r_{ij} when it is firing at them.

$e_{ij}(r)$ = *the allocation factor.* The proportion of the ith Blue group systems assigned to fire on the jth-group Red targets which are at range r_{ij}.

$I_{ij}(r)$ = *the intelligence factor.* The proportion of the ith group firing Blue weapons allocated to the jth Red group which are actually engaging live j^{th}-group Red targets at range r_{ij}.

$F_{ij}(t)$ = *the terrain/acquisition factor.* The percentage of ith group survivors who have detected a target in the jth target group at time t.

Similar definitions exist for the components of the Red attrition coefficient, B_{ji}. In section 3.1 of this paper we defined the attrition rate at a

---

[1]Research done on comparing the deterministic and stochastic formulations for the homogeneous-force case (only one force group on each side) indicates that the deterministic formulations are reasonably good approximations to the expected number of survivors if there is a small probability that either side is annihilated. Additionally, it is noted that in many defense studies that employ Monte Carlo simulations, only the expected results are considered in the decision-making process.

particular range[1] to target as

$$\alpha_{ij}(r_{ij}) = \frac{1}{E[T_{ij}|r_{ij}]}$$

and described a model for predicting the rate for a tank gun system as a function of its various performance capabilities. Similar models have been developed for other types of weapon systems and firing doctrines. Methods developed to predict the other components of the model are described in succeeding subsections.

The Allocation Factor

As noted earlier, the allocation factor is the proportion of the ith Blue group systems assigned to fire on j^{th}-group Red targets. This is included since only those systems directing their fire (or other lethal effects) on the jth group or its area are likely to cause attrition of the target. The allocation factor may be input by military judgment reflecting the assignment strategies deemed most appropriate to the tactical situation. This factor may be input directly or determined from a priority or target worth scheme.

In addition to the use of military judgment to assign weapon groups to target groups (the procedure usually employed in the model when used by the Army), research results on allocation strategies using differential game concepts (Isaacs, 1965) have given rise to an approximate optimal[2] priority ordering rule which is used in some simpler versions of the model. The rule is

"Blue weapon group i should engage live Red target group K if the product

$$\alpha_{iK}\beta_{Ki} > \alpha_{ij}\beta_{j} \qquad \text{for all } j$$

and Red weapon group j should engage live Blue target group K if the product

$$\beta_{jK}\alpha_{Kj} > \beta_{ji}\alpha_{ij} \qquad \text{for all } i."$$

This rule is used over all eligible targets, which consist of those targets within range which are not externally prohibited.[3] If all eligible target groups are unable to return fire, then all of the above products will be zero.

---

[1]For clarity of discussion, variations in the attrition rate due to changes in target posture, environmental effect, etc. which are included in the model are not presented in this section.

[2]Optimal with respect to linear end of battle measures such as the difference (Blue minus Red) in survivors.

[3]Externally prohibited targets for a weapon group are those for which the attrition rate is zero, e.g., a rifle against a tank target.

In this case, Blue group i is assigned to fire on that target group j for which α_{ij} is maximum and an analogous rule is used for assignment of a Red weapon group.

The Intelligence Factor

As noted earlier, the intelligence factor is the proportion of the ith group firing Blue weapons allocated to the jth Red group which are actually engaging live j^{th}-group Red targets. This factor is included to consider the loss in efficiency (effectiveness) of a firing weapon when it is firing on either targets already attrited or on areas that are void of targets. Based on some renewal theory modeling, the intelligence factor is predicted as

$$I_{ij}(r_{ij}) = \frac{p_L \overline{T}_L}{p_L \overline{T}_L + p_D \overline{T}_D + p_V \overline{T}_V} , \tag{47}$$

where

p_L = the probability of firing on a live target, given fire on a target,

p_D = the probability of firing on a dead target, given fire on a target,

p_V = the probability of firing on a void area, given fire on a believed target,

$\overline{T}_L$ = the expected or average time to fire on a live target before switching fire (given by equation 40),

$\overline{T}_D$ = the expected or average time to fire on a dead target before switching fire, and

$\overline{T}_V$ = the expected or average time to fire on a void area before switching fire.

Fire is directed at attritted targets or voids (false targets) because of the uncertainties associated with detecting, recognizing, and identifying real targets under the stress of battle. The intelligence factor dependence on range is due to both the probabilities and the times in equation (47).

Computation and Terrain Interactions

As implied throughout this paper, there exist a number of operating versions of the differential ground combat models varying in degrees of complexity and simplifying assumptions. At one end of the continuum are simplified models (homogeneous forces, constant attrition-coefficient heterogeneous forces, no terrain effects, etc.) which succumb to closed-form mathematical solution techniques. These are used principally for theoretical research. At the other end of the continuum are more realistic complex versions that include more terrain effects and larger dimensionality in the input data (i.e., kill probabilities, which depend on the target type, cover status, movement status, aspect angle, etc.). These models usually require the use of numerical solution procedures and are used principally for analysis in weapon system studies. Prior to discussing the terrain/acquisition factor, I shall briefly summarize the computational procedure employed and the way that terrain effects are included in one of the versions of the model.

In the computational program, the basic differential equations are approximated by the difference equations

$$m_i(t + \Delta t) = \max \left\{ 0,\ m_i(t) - \sum_{j=1}^{J} B_{ji}(t) n_j(t) \Delta t \right\} \quad \text{for } i = 1, 2, \ldots, I$$

$$n_j(t + \Delta t) = \max \left\{ 0,\ n_j(t) - \sum_{i=1}^{I} A_{ij}(t) m_i(t) \Delta t \right\} \quad \text{for } j = 1, 2, \ldots, J,$$

where Δt is the computational time step. A 10-second time step is usually used. In applications of the model to battalion-level task force engagements with weapon systems currently under study it was observed that smaller time steps did not alter the solution, while larger steps led to significant errors (overkills, failure to switch assignments, etc.). Clearly, the time step must be appropriately selected depending on the capabilities of the systems involved and the scenario activity.

The correspondence between battle time and the spatial distribution of forces during the battle is obtained from knowledge of predetermined movement patterns of all Red and Blue groups which are input to the model. These movement patterns are obtained from a terrain preprocessor. Routes of advance and movement tactics (sections leapfrog, sections advance and provide covering fire, etc.) are input to the preprocessor, which then considers the terrain characteristics (soil type, roughness, grade, etc.) along the routes and maneuver capabilities of the weapon systems (speeds, accelerations, etc.) to generate the time-sequenced movement patterns.

The effects of terrain which limit the firing activity due to loss of acquisition capability are also included in this model. For this purpose the terrain is incorporated in this version of the model as if it were a map with digitized properties of concealment, cover (line-of-sight), etc. associated with each location or pairs of locations. The model considers these terrain effects (which are also obtained from the preprocessor) in developing the terrain/acquisition factor.

Terrain/Acquisition Factor

The terrain/acquisition model considers that the acquisition process occurs *in parallel* with the firing and movement processes. Since the overall model considers *groups* of weapon systems, the parallel acquisition submodel is designed to determine the percentage of observers in a group who have detected a target in an opposing group. Detections occur due either to visual sighting of nonfiring targets or pinpointing the flash of a firing target. The terrain effects of visibility (fully exposed, partially exposed, not exposed) and line-of-sight[1] (exists, does not exist) are interfaced with the visual detection and pinpointing capabilities by the following computational formulas:

$\overline{F}_{ij}(t + \Delta t)$ = percentage of ith-group survivors who have *failed* to detect a target in j^{th} target group at time $(t + \Delta t)$

$$= \begin{cases} 1 & \text{if no LOS} \\ \overline{F}_{ij}(t)\overline{V}_{ij}(t, t + \Delta t)\overline{P}_{ij}(t, t + \Delta t) & \text{if LOS exists,} \end{cases} \quad (48)$$

where

$\overline{V}_{ij}(t, t + \Delta t)$ = percentage of ith-group survivors who have *failed* to visually detect a target in j^{th} target group in the interval $(t, t + \Delta t)$

$$= \begin{cases} 1 & \text{if visibility does not exist} \\ e^{-\lambda_{ij}(t+\Delta t)\cdot\Delta t\cdot n_j(t)} & \text{if visibility exists} \end{cases} \quad (49)$$

$\overline{P}_{ij}(t, t + \Delta t)$ = percentage of ith-group survivors who have *failed* to pinpoint a target in j^{th} target group in the interval $(t, t + \Delta t)$

$$= \begin{cases} 1 & \text{if no LOS} \\ (1 - p_{ij})^{\frac{\rho_{ij}(t,t+\Delta t)}{n_j(t)} n_j(t)} & \text{if LOS exists} \end{cases} \quad (50)$$

and

$\lambda_{ij}(t)$ = the detection rate at time t, which is a function of range between the ith and j^{th} groups and exposure;

$n_j(t)$ = the number of surviving j^{th}-group targets at time t;

---

[1]Abbreviated LOS.

p_{ij} = probability of ith-type observer pinpointing a jth-type weapon when it fires *one round*. This is considered dependent on range between i^{th} and j^{th} groups, the weapon type, observer type, and movement status;

$\rho_{ij}(t, t + \Delta t)$ = the number of rounds fired by the jth target group in the interval $(t, t + \Delta t)$.

The formulas assume a Poisson visual detection process when visibility exists and a geometric pinpointing process when LOS exists.

The percentage of ith-group survivors *who have detected* a target in the jth target group at time $(t + \Delta t)$ is obtained directly from $\overline{F}_{ij}(t + \Delta t)$ for all j. Using this information, and an input military worth or target priority rule, appropriate numbers of the ith-group survivors are assigned to fire on different target groups.

Model Output

Output of this version of the model at each time step is a status listing of all weapon groups which are involved in firing events (firer or target). The listing contains the group numbers, their locations, range separations, movement status, visibility status, percentage of the firing group allocated, round-type used, attrition rate, and amount of attrition. A summary output is also provided each 10-second time step. This summary lists the cumulative number of losses of each weapon *type* by the *types* in the opposing force which caused the attrition.

Comparisons with Monte Carlo Simulations

When initially developed, the defense community was very reluctant to use this hybrid analytic/simulation model because of its belief that only detailed Monte Carlo simulations with their focus on individual systems and activities could credibly represent the complexity of a land battle. For this reason the model was "tested" by comparing its combat predictions to the results predicted by more detailed Monte Carlo simulations. A few of the comparison results are presented in this subsection of the notes.

Figure 5 depicts one of the tactical plans considered in one of the comparison studies. The tactical plan shown is a Blue attack engagement against a fixed Red defensive position. The attack is conducted along three major axes with four individual routes of advance per axis. Each route consists of individual main battle tanks and/or supporting armored personnel

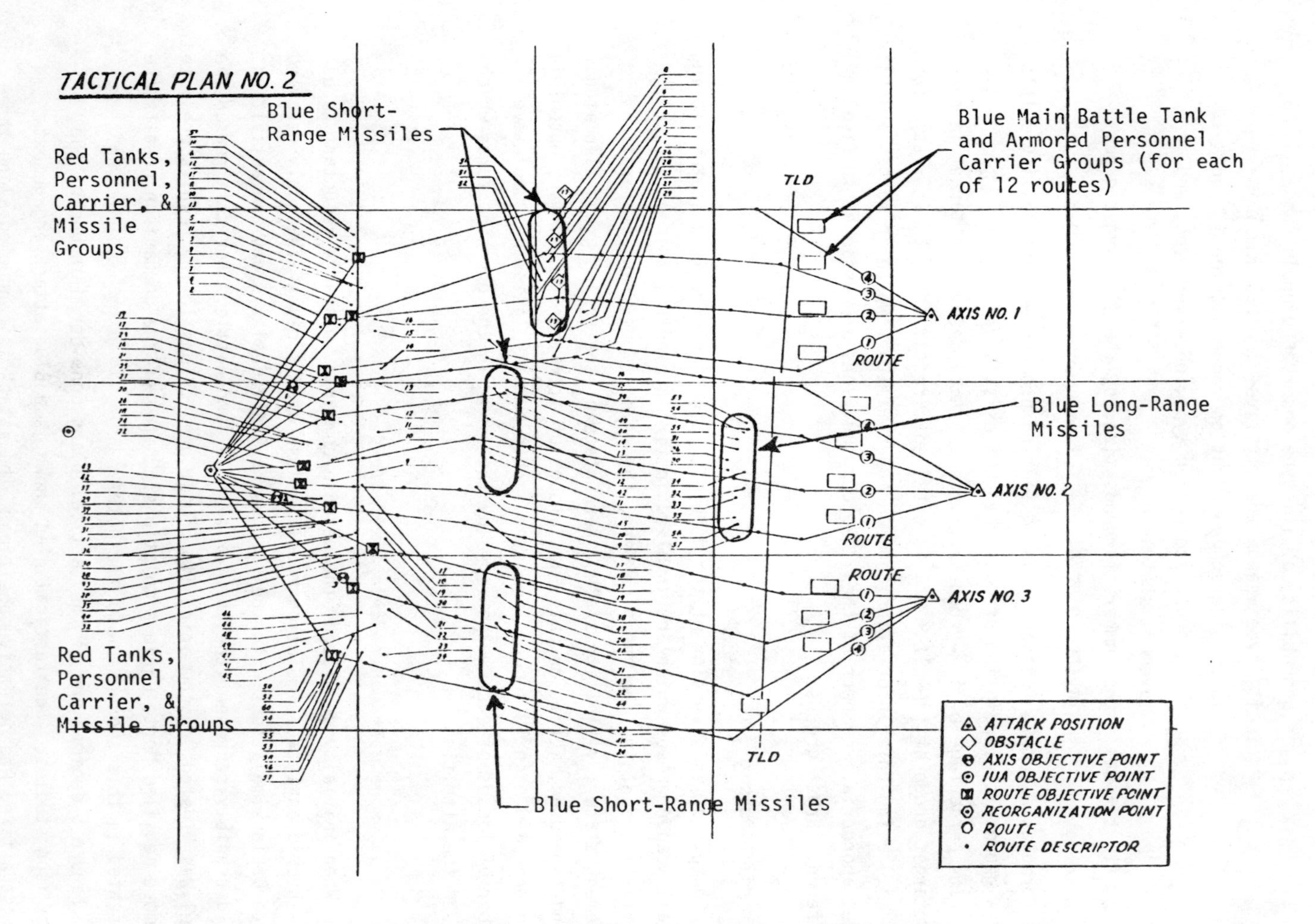

FIGURE 5 TACTICAL PLAN FOR BLUE ATTACK ENGAGEMENT

carriers equipped with rapid-fire weapon systems. In addition to these maneuver units of main battle tanks and personnel carriers, the Blue attack force had long-range missiles and short-range missiles, as shown in the figure. The defending force is comprised of tanks, missiles, and armored personnel carriers equipped with rapid-fire weapons systems.

The Monte Carlo simulation of this engagement considered the movement, acquisition, and combat activity (duels) of each and every element in the battle.[1] Maneuvers, in terms of attack speed and accelerations, over different portions of the terrain were considered for each weapon based on preprocessed terrain analysis. The existence or nonexistence of line-of-sight between weapons systems for each route to all other weapon systems was used as input. Preprogrammed target priority tables were used to specify the allocation of individual weapons to targets. A replication of the simulation consisted of moving each of the systems down their prespecified paths and evaluating, by Monte Carlo means, the acquisition and attrition processes (the fundamental duel event) for each weapon system during the course of the engagement. The engagement was replicated many times to obtain a level of statistical stability for the results.

The hybrid combat model was applied to this and other engagements by aggregating individual weapons systems into groups. Thus, for each route on an axis there were two separate groups of main battle tanks or armored personnel carriers (APCs). The long-range missiles were aggregated into one group and the short-range missiles were aggregated into three groups, one for each axis. The Red defensive force was aggregated by weapon type for each axis, thus producing nine Red defensive groups. Also included, but not shown in the figure, were indirect-fire artillery weapons systems for both forces.

Using the attrition-rate models discussed in section 3.1, the attrition rates for each group on appropriate target groups were calculated using the same basic firing time, accuracy, and lethality data used in the simulation. Target acquisitions were determined with the parallel acquisition model using the same detection rate and pinpoint probabilities used in the simulation. The allocation factors (e_{ij} and h_{ji}) employed were based on the priority tables used in the simulation. The intelligence factor was set equal to 1.0 since these effects were not considered in the simulation. Mobility and line-of-sight data from the preprocessor were considered in a deterministic manner similar to that employed in the simulation. Average speeds and line-of-sights over segments of the routes were input for each of the *aggregated* groups.

---

[1]Some of the engagements considered as many as 100 individual weapon systems.

Thus, a group was moved as a whole, and visibility did or did not exist to the group as an entity. The differential equations were solved numerically using computational time-step procedures.

Using this approach, the model was applied initially to short-range defense and long-range attack scenarios considered in a specific study program. With these engagement types, runs involving different weapon systems and force structures were made for comparison with the simulation results. Some of these comparisons are shown below in tables 1 and 2.

Table 1 presents a comparison of the results of one of the short-range defense engagements. The initial numbers of forces and the numbers of survivors at three analysis points as predicted by both Monte Carlo simulation and the hybrid model are given. The analysis points are defined by the percentage of Red tank survivors: low equal to 70 percent, principal equal to 50 percent, and high approximately equal to 20 percent. The times at which these analysis points are reached in each of the models also is given. Two sets of results at the low analysis point in the hybrid model are shown since there was an appreciable attrition in the 240-250 time interval.

Table 2 presents the comparisons of tank survivors only at the three analysis points for another short-range defense and one long-range attack engagement. Table 3 presents comparisons of six pure tank battles in terms of their *loss exchange ratios* (LER) at the end of the battle. The LER is the ratio of enemy (Red) to friendly (Blue) armored system losses and is interpreted as the expected trading ratio between two specific combatant forces if they were to engage in many similar battles.

The initial favorable comparisons between this hybrid analytic/simulation of battalion level combat activities and the detailed Monte Carlo simulation strongly supported the hypothesis that both models were essentially describing the same combat process. This, and hundreds of additional comparisons since then, have given credence to this model and its many successive versions developed over the past ten years. These models have been used extensively during this period, providing information to address many system requirements and system choice issues. Additionally, the models have provided some useful insights regarding tactics for the defense of Western Europe.

Current estimates are that the Warsaw Pact force could attack the NATO Alliance along the West German border with armored force ratios ranging from 3:1 to 6:1. Analysis of many small unit actions indicates that in a single battalion- or company-sized defensive engagement, the instantaneous *loss exchange ratio*[1] as a function of battle time is as depicted in figure 6.

---

[1]Mathematically, the ratio of the rates of attacker and defender losses.

TABLE 1 COMPARISON OF SURVIVING FORCES

Short-Range Defense

Initial Numbers

16 Blue Tanks	40 Red Tanks
6 Blue Short-Range Missiles	0 Red Missiles
6 Blue APC	12 Red APC
3 Blue Long-Range Missiles	

ANALYSIS POINT	WEAPON	SIMULATION	TIME	HYBRID ANALYTIC/ SIMULATION	TIME
	Blue Tanks	13.90		15.1/13.9	
	Blue SR Missiles	5.10		6.0^-	
Low	Blue APC	5.93	242	6.0^-	240/250
	Blue LR Missiles	2.73		3.0^-	
(70%)	Red Tanks	28.00		30.4/24.4	
	Red Missiles	-----		-------	
	Red APC	11.70		11.0/10.6	
	Blue Tanks	12.23		12.6	
	Blue SR Missiles	4.57		6.0^-	
Princi-	Blue APC	5.73	263	6.0^-	260
pal	Blue LR Missiles	2.27		3.0^-	
(50%)	Red Tanks	20.00		19.2	
	Red Missiles	-----		-------	
	Red APC	10.33		10.2	
	Blue Tanks	9.40		10.0	
	Blue SR Missiles	2.97		5.8	
High	Blue APC	5.20	327	6.0^-	290
	Blue LR Missiles	2.00		2.9	
(22%)	Red Tanks	8.90		7.2	
	Red Missiles	-----		-------	
	Red APC	4.27		7.0	

TABLE 2 COMPARISON OF SURVIVING TANKS

BLUE FORCE MISSION	WEAPON	INITIAL NO.	LAP		PAP		HAP	
			SIM.	HYBRID	SIM.	HYBRID	SIM.	HYBRID
Short-Range Defense	(Time)		(242)	(240)	(259)	(260)	(352?)	(280)
	Blue Tank	19	17.20	18.0	15.87	15.4	13.33	13.7
	Red Tank	40	28.00	30.4	20.00	17.6	8.0	8.8
Long-Range Attack	(Time)		(206)	(260)	(411)	(440)	(512)	(470)
	Blue Tank	31	26.30	27.6	23.83	23.8	21.47	23.5
	Red Tank	13	9.00	9.1	7.0	7.0	2.0	1.4

TABLE 3 COMPARISON OF LOSS EXCHANGE RATIO FOR PURE TANK BATTLES

		Initial Number of Tanks			Loss Exchange Ratio at End of Battle	
		Blue	Red	Ratio Red/ Blue	Simulation	Hybrid Analytic/ Simulation
ATTACK	Short Range	54	40	.74	1.08	.92
	Medium Range	54	40	.74	1.10	1.10
	Long Range	54	27	.50	.72	.96
DEFENSE	Short Range	46	72	1.57	2.10	2.10
	Medium Range	48	72	1.50	1.95	1.68
	Long Range	38	72	1.89	2.00	1.87

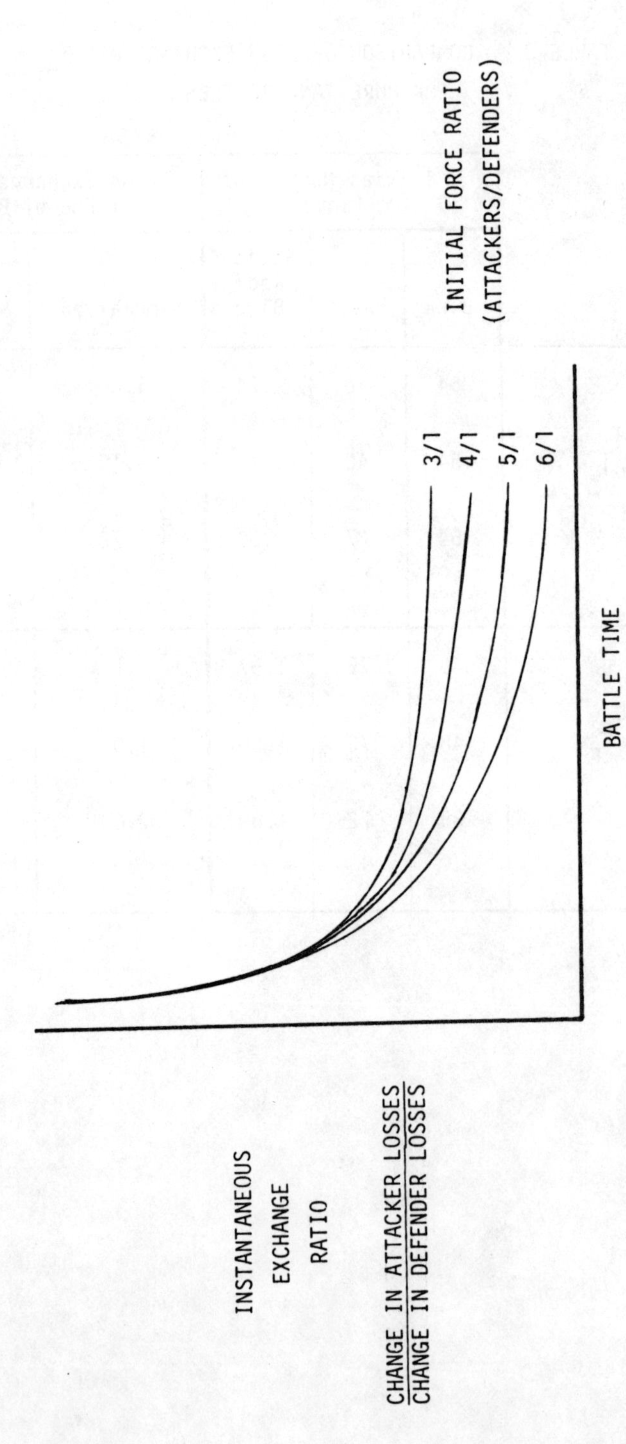

FIGURE 6 INSTANTANEOUS EXCHANGE RATIO AS A FUNCTION OF BATTLE TIME

The instantaneous exchange ratio is very high and relatively independent of the force ratio (and particularly of threat size) early in the battle because of concealment and first shot advantages accrued the defender. The instantaneous exchange ratio advantage moves to the attacker as the forces become decisively engaged, because more attackers find and engage targets and the concentration and saturation phenomena come into play for the attacker.

In terrains that permit it, this suggests that an in-depth use of a large number of company and below, brief, direct fire engagements or ambushes may be an effective tactic. Such a campaign might occur in two phases.

Phase I: This phase involves a somewhat constant density of small mechanized antitank (AT) teams who engage threat forces in carefully selected (and perhaps prepared) terrain locations. They operate at the very high, essentially force ratio independent, part of the instantaneous exchange ratio curves by firing a small number of rounds and immediately withdrawing before suffering any attrition. The withdrawal is made to another engagement site, passing through a similar site occupied by another AT team.

Phase II: This phase continues the sequence of engagements of phase I; however, increasing amounts of enemy attrition are to be obtained per unit of ground traded at the expense of defender losses. This is achieved by increasing the density of AT teams or by organizing increasingly larger combat units who engage for longer periods of time (thus moving farther down the instantaneous exchange ratio curve). This sequence of progressively hardening the defense continues until the final boundary line at which time units are organized for the "decisive battle" if necessary.

Rather than focus on obtaining good force ratios for individual engagements as do other tactical concepts, this tactical concept attempts to make use of conjectured, early-on, high, instantaneous exchange ratios. A number of weapon system acquisitions and tactical changes in the European theater are based on these concepts.

4.0 OVERVIEW OF CURRENT STATUS AND DEVELOPMENT TRENDS

As previously noted, utility of the initial differential hybrid analytic/ simulation model of small unit engagements gave rise to a number of variants in the late sixties and seventies. The period 1972-1978 produced additional improvements in modeling combat activities. Using some improved modeling techniques,[1] many of the existing models were improved and new models developed for the three levels of battles over the past eight years. The directions of these improvements and associated trends are shown in figure 7. Although high resolution Monte Carlo simulations are still used at the battalion level, a large number of hybrid analytic/simulation models have been developed and are extensively used in tactical warfare studies of system design requirements, system choice, and system mix issues. Using the same kinds of input data, these models have been shown to generate essentially the same battle results as the simulation models but orders of magnitude more efficiently. At the division-corps level, a number of the war games have been improved to reduce player participation, and, more recently, a number of hybrid analytic/simulation models have been developed. At the theater level, simplified "firepower score" analytic structures have been replaced by first simple and then more complex hybrid analytic/simulations. Finally, over the last year or two, some initial analytic techniques for aggregating and extrapolating results of the more dynamic hybrid analytic/simulation models and war games at the corps and theater level have been developed.

It is interesting to note the trends that appear to have occurred at each of the levels of tactical warfare modeling. At the battalion level the models have progressed from simple Monte Carlo and analytic duels to very detailed simulatory structures and appear to be moving more in the analytic direction with complex hybrid analytic/simulations. At the division-corps level the progression has been from simplified manual and computer-assisted war games to high-resolution detailed games, and then to detailed hybrid analytic-simulations. At the theater level the progression has been from simplified war games, to simplified analytic models, to much more complex high resolution hybrid analytic/simulations. Considering all of the developments over the twenty-some odd years, the trend is similar to that observed in a number of other modeling areas: initial developments are rather simplified, mostly analytic models which progress to very detailed high resolution structures that are somewhat simulatory in nature, and then more sophisticated analytic structures are used to describe the complexity and detail. Accordingly it is my impression that in the next decade the developments in each of the three areas will move in the direction shown on figure 7. I believe that an essentially pure analytic model of battalion level activities will be developed,

---

[1]The interested reader is referred to Bonder (1978) for an overview description of advances in modeling technology which occurred in this period.

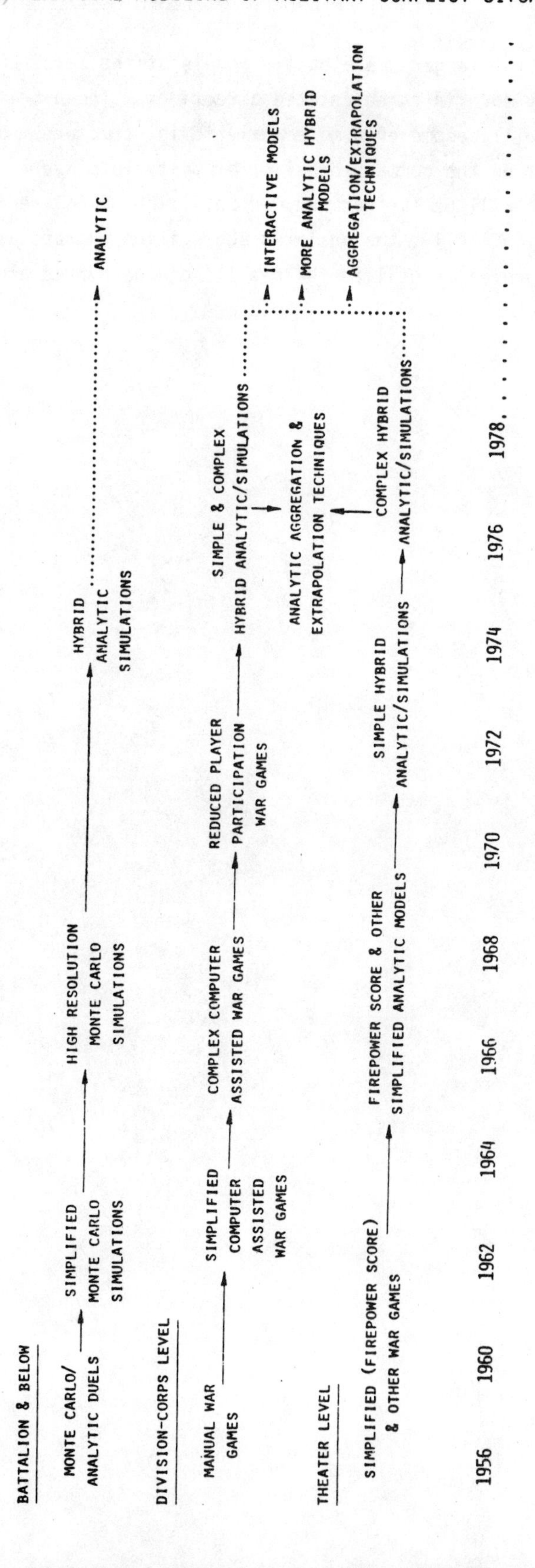

FIGURE 7 TRENDS IN TACTICAL WARFARE MODEL DEVELOPMENTS

possibly by 1985. The larger-scale battle models at the corps and theater level will probably move in three related directions. The hybrid analytic/simulation models will become even more analytic in structure. Because of the difficulty in modeling the command behavior of multiple players, player participation in war games will be further reduced until the games are in essence interactive models. Finally, the analytic aggregation/extrapolation techniques will be further improved to utilize the results of the hybrid and interactive models.

5.0 BIBLIOGRAPHY

Adams, H.E., et al., "Carmonette: A Computer-Played Combat Simulation," Technical Memorandum ORO-T-389, Operations Research Office, The Johns Hopkins University, February 1961.

Ancker, C.J., and Gafarian, A.V., "The Distribution of Rounds Fired in Stochastic Duels," Naval Research Logistics Programming 11, 4 (1964) 303-327.

Ancker, C.J., Jr., and Williams, T., "Some Discrete Processes in the Theory of Stochastic Duels," Operations Research 13, 2 (1965) 202-216.

Ancker, C.J., Jr., and Gafarian, A.V., "The Distribution of the Time Duration of Stochastic Duels," Naval Research Logistics Quarterly 12, 3 and 4 (1965) 275-294.

Ancker, C.J., Jr., "Stochastic Duels of Limited Time Duration," Canadian Operational Research Society Journal 4, 2 (1966a) 59-81.

Ancker, C.J., Jr., "The Status of the Development of the Theory of Stochastic Duels-II," SP-1017/008/01, System Development Corporation, Santa Monica, CA, September (1966b).

Bach, R.E., Dolansky, L., and Stubbs, H.L., "Some Recent Contributions to the Lanchester Theory of Combat," Operations Research 10, 3 (1962) 314-326.

Barfoot, C.B., "The Lanchester Attrition Rate Coefficient: Some Comments on Seth Bonder's Paper and a Suggested Alternate Method," Operations Research 17, 5 (1969) 888-894.

Benjamin, W.C., and Gholston, W., Compilation of Results Obtained from Some U.S. Firings of Kinetic Energy Projectiles against Tanks (U), Confidential, Ballistics Research Laboratories Memorandum Report No. 1295, August 1960.

Bonder, S., "The Lanchester Attrition-Rate Coefficient," Operations Research 15, 2 (1967) 221-232.

Bonder, S., "Mathematical Models of Combat," in Topics in Military Operations Research, The University of Michigan Engineering Summer Conferences, 21 July-1 August 1969.

Bonder, S., "The Mean Lanchester Attrition Rate," Operations Research 18, 1 (1970) 14-23.

Bonder, S., "Perspectives from Defense Modeling," keynote address presented at Washington Operations Research Council Symposium, Models in Government: New Developments and Applications, Washington, DC, 20 October 1978.

Bonder, S., and Farrell, R.L., (Eds.), Development of Models for Defense Systems Planning, Report No. SRL 2147, TR 70-2, Systems Research Laboratory, Department of Industrial Engineering, The University of Michigan, September 1970 (National Technical Information Service, Springfield, Virginia, Report No. AD 715 664).

Bowen, K.C., "Analytical Models for the Study of Battle Outcome Distributions," Third Tripartite Naval Operational Research Symposium, March 1963.

Brackney, H., "The Dynamics of Military Combat," Operations Research 7, 1 (1959) 30-44.

Brown, R.A., "A Validation Study of Certain Combat Models," Unpublished Working Paper, Systems Research Group, The Ohio State University, Columbus, Ohio, (1959).

Brown, R.H., A Stochastic Analysis of Lanchester's Theory of Combat, ORO-T-323 (U), Operations Research Office, The Johns Hopkins University, Chevy Chase, MD (1955).

Brown, R.H., "Theory of Combat: The Probability of Winning," Operations Research 11 (1963) 418-425.

Clark, G., "The Combat Analysis Model," Ph.D. Dissertation, Department of Industrial Engineering, The Ohio State University, (1968).

Clarke, B.C., "The Offensive Employment of Tanks," Armor LXXI 3 (1962) 42-43.

Combat Developments Command, Armor Agency, U.S. Army, "Tank, Antitank, and Assault Weapons Requirements Study," Fort Knox, KY: USACDC Armor Agency (1969).

Cramer, H., Mathematical Methods of Statistics, Princeton University Press, Princeton, NJ (1961).

Deitchman, S.J., "A Lanchester Model of Guerrilla Warfare," Technical Note 62-58, Institute for Defense Analysis, October 1962.

Dolanský, L., "Present State of the Lanchester Theory of Combat," Operations Research 12 (1964) 344-358.

Dresher, Melvin, Games of Strategy - Theory and Application, Prentice-Hall, Englewood Cliffs, NJ (1961).

Engel, J.H., "A Verification of Lanchester's Law," Operations Research 2 (1954) 163-171.

Evans, G.W. II, Wallace, G.F., and Sutherland, G.L., Simulation Using Digital Computers, Prentice-Hall, Inc., Englewood Cliffs, NJ (1967), Chapter 4.

Freedman, R., "Vulnerability Procedures for Direct-Fire Projectiles," Chapter 3 in The Tank Weapon System, Report No. RF 573 AR 65-1 (S) Systems Research Group, The Ohio State University, Columbus, Ohio (1965).

Feller, W., An Introduction to Probability Theory and Its Application, John Wiley and Sons, Inc., New York, NY (1957).

Gause, A., "Command Techniques Employed by Field Marshall Rommel in Africa," Armor LXVII 4 (1958) 22-25.

Gnedenko, B.V., The Theory of Probability, Chelsea Publishing Company, New York, NY (1962).

Helmbold, R.L., "Some Observations on the Use of Lanchester's Theory for Prediction," Operations Research 12 (1964) 778-781.

Helmbold, R.L., "A 'Universal' Attrition Model," Operations Research 14, 4 (1966) 626-635.

Hogg, R. V., and Craig, A.T., Introduction to Mathematical Statistics, Macmilliam, New York, NY (1959).

Isaacs, R., Differential Games, John Wiley and Sons, Inc., New York, NY (1965).

Kimball, G.F., and Morse, P.M., Methods of Operations Research, John Wiley and Sons, New York, NY (1951).

Lanchester, F.W., Aircraft in Warfare: The Dawn of the Fourth Arm - Vol. V., The Principal of Concentration, Engineering 98 (1914) 422-423.

Marshall, C.W., "Probabilistic Models in the Theory of Combat," Transactions of the New York Academy of Sciences, Ser. II, 27, 5 (1965) 477-487.

Montross, L. and Canzona, N.A., U.S. Marine Operations in Korea, 1950-1953, II, Washington (1955).

O'Ballance, E., The Sinai Campaign of 1956, Fredrick A. Praeger, New York, NY (1959).

Parzen, E., Stochastic Processes, Holden Day Inc., San Francisco, CA (1962).

Rapoport, A., "Lewis F. Richardson's Mathematical Theory of War," Conflict Resolution 1, 3 (1957) 249-299.

Richardson, L.F., Arms and Insecurity, Quadrangle Books, Inc., Chicago, IL (1960a).

Richardson, L.F., Statistics of Deadly Quarrels, Quadrangle Books, Inc., Chicago, IL (1960b).

Robison, S.S. and Robison, Mary L., A History of Naval Tactics from 1530 to 1930, U.S. Naval Institute, Annapolis, MD (1942).

Saaty, T.L., Mathematical Methods of Operations Research, McGraw-Hill Book Co., New York, NY (1959).

Schaffer, M. B., "Lanchester Models of Guerrilla Engagements," Operations Research 16, 3 (1968) 457-488.

Schoderbek, J.J., "Some Weapon System Survival Probability Models - I. Fixed Time Between Firings," Operations Research 10, 2 (1962) 155-167.

Snow, R., Contributions to Lanchester's Attrition Theory, Memorandum RA-15078, RAND Corporation, April 1948.

Taylor, J. and Brown, G., "Canonical Methods in the Solution of Variable-Coefficient Lanchester-Type Equations of Modern Warfare," Operations Research 24 (1976), 44-69.

Vector Research, Incorporated, "Modification and Improvement of Differential Models of Combat," VRI-2 FR 70-1(U), Vols. I and II, Vector Research, Incorporated, Ann Arbor, Michigan, October 1970.

Wagner, H., Principles of Operations Research: With Applications to Managerial Decisions, Prentice-Hall, Inc., Englewood Cliffs, NJ (1969).

Weiss, H. K., "Lanchester-Type Models of Warfare," Proceedings of the First International Conference on Operational Research," Operations Research Society of America, Baltimore, MD (1957) 82-98.

Weiss, H.K., "The Fiske Model of Warfare," Operations Research 10 (1962) 569-571.

Weiss, H.K., "Review of Lanchester Models of Warfare," Paper presented at 30th National Meeting of the Operations Research Society of America, Dunham, North Carolina, October 1966.

Whitney, D.R., Elements of Mathematical Statistics, Henry Holt and Co., New York, NY (1959).

Willard, D., Lanchester as Force in History: An Analysis of Land Battles of the Years 1618-1905, RAC-TP-74, Research Analysis Corporation, November 1962.

Williams, T., "Stochastic Duels - II," SP-1017/003/00, Systems Development Corporation, Santa Monica, CA, September 1963.

Williams, T. and Ancker, C.J., "Stochastic Duels," Operations Research 11, 5 (1963) 803-817.

Zeller, G.A., Methods of Analysis of Terminal Effects of Projectiles against Tanks (U), Secret, Ballistics Research Laboratories Memorandum Report No. 1342, April 1961.

VECTOR RESEARCH, INCORPORATED
PO BOX 1506
ANN ARBOR, MICHIGAN 48106

Proceedings of Symposia in Applied Mathematics
Volume 25, 1981

QUEUEING NETWORKS

Ralph L. Disney[1]

ABSTRACT. The study of queueing networks is a rather recent addition to queueing theory. In this paper we will review about the last 20 years of developments in the area. Primary emphasis is placed on Jackson networks and the major contribution of Kelly. We discuss queue length processes, waiting time processes, busy period processes, and departure processes. We will also note a few new studies that have appeared in the area of network flows. Topics that are not discussed in detail in the paper are briefly noted in the final section. The bibliography can serve as an introduction for further reading in the area.

1. Introduction and Some Background

1.0 Introduction. In this paper we will briefly review some developments that have occurred over the last 20 years, in an area of applied mathematics called queueing network theory. The need for such applied work has been clear almost since the beginning of what is called queueing theory. Some of the work of A. K. Erlang (see Brockmeyer, et al. [1948]) seems to be pointing toward a study of interconnected systems of service centers. Certainly by the 1930's it was clear that such work was evolving. (See chapters 7-10 in Syski [1960].) In most of these early studies, the developments appear to be closely tied to attempts to solve problems that occur principally in telephone switching systems.

1980 Mathematics Subject Classification 60 K25, 90B22.

[1]This research was supported jointly by NSF Grant ENG77-22757 and by the Office of Naval Research Contract N00014-77-C-0743 (NR042-296). Distribution of this document is unlimited. Reproduction in whole or in part is permitted for any purpose of the United States Government.

Much of classical queueing theory (up to about 1957 or 1960) was concerned with properties of Markov processes and except for names given to the parameters and processes it is difficult to separate out the peculiarities of queues from other Markov processes that were being used as models, for example, in biological modelling. In fact, many of the birth-death models of biological studies have interesting and useful counterparts in queueing theory. For this reason a carefully drawn history of queueing theory would be strained because of the interrelations with other fields. The useful thing to remember is that in its beginning phases the field was closely allied with attempts to solve real life problems.

The 1950's were perhaps the "golden age" of developments in the field. Under many impacts (including the introduction of more sophisticated and realistic models into problems that heretofore had been the domain of the industrial engineer) the field began to take on a life of its own. In much of the 1950-60 work and extending even to today, the work in the field veered more and more away from its applied bent. It was at this time that queueing theory started to be sufficiently arcane that many researchers were turned away from it. Easy queueing problems had been solved. Many papers were published on special cases of birth-death processes. The field was moribund after about 1965-70.

Starting about 1965, the field of computer systems analysis began to delve into properties of time sharing and interconnected systems. Here the researchers encountered many of the same types of problems encountered earlier in telephone systems and production systems. It was natural, therefore, to use existing, known results. By 1970 and certainly by 1973 research into computers and computer systems had uncovered a large number of problems beyond the work in earlier queueing theory.

At the same time that these developments in computer systems were taking place, new problems were occurring and old unsolved problems were becoming more pressing in telecommunication theory and production theory. New areas of application were evolving in military command and control modelling;

disease modelling, military tactics modelling, public sector models for police, fire, and medical emergency systems design; and many others. All of these topics put demands on queueing theory (and indeed other areas of probability and random processes) which that theory was unable to handle. Systems (e.g., networks) of queues rather than single queues became areas of important study in many of the above applications. Markov process theory did not provide a rich enough body of knowledge so new ideas evolved.

Largely because the applied topics put demands on existing theory that could not be accommodated quickly enough, much of the applied work turned to Monte Carlo simulation of systems. But that placed a new load on the theory of statistics as well as probability, random processes, and queueing. Because of the relative quiescence of the 1957-70 period, queueing theory lost ground to applied topics.

1.1 Purpose. The purpose of this paper is to briefly summarize where we stand in the study of queueing network theory. Time and space do not permit any in-depth discussion here. The bibliography should be consulted by the reader interested in "reading themselves into" the area. However, that reader should be forewarned that papers on these topics are spread over a large number of journals throughout the world. There is no one best place to look for research papers or to expect to follow diligently and thereby stay up with the field. For background into queueing theory the reader might consult Syski [1960] who does a masterful job summarizing work in queueing theory up to about 1960. Kleinrock's two volumes [1975; 1976] summarize many known results in queueing theory and give a nice discussion of many of the problems of queueing phenomena occurring in computer systems. One can find some discussions concerning section 2, to follow, in most post 1975 texts. Material in section 4 is nicely presented in Kelly [1979]. Topics in section 5 and 6 have not been pulled together in any one place to the best of our knowledge so the reader is on his own there. References are provided herein but that list is far from complete.

1.2 Some Background, Notation and Symbolism. In order to embed our future discussion in the more classic field of queueing theory, we will present a brief review of a few ideas in basic queueing theory.

At an intuitive level queueing theory is concerned with problems arising in waiting lines. One supposes that there is something that can perform a needed service, a server. In the simpler cases it is assumed that the sequence of service times is a sequence of mutually independent random variables that are identically and non-negatively distributed. In queueing theory such a service process is called a <u>G-type service process</u>. In the theory of stochastic processes such a process is called a <u>renewal process</u>.

The demand for this service is usually modelled with the times between demands being the variables of interest. These times are assumed to be a sequence of independent, identically distributed random variables that are non-negative (another renewal process). Such an arrival process is called a <u>GI-type arrival process</u>. It is usual to assume that the arrival process and service time process are independent processes. Such a queueing system is called a GI/G/1 queue. The last argument specifies the number of servers in the system.

Queueing or waiting occurs in such systems whenever an arrival occurs to find the server already engaged serving a previous arrival. If an arrival occurs and the server is not so engaged, that arrival immediately enters service, under the usual assumptions. (There are studies that do not include this assumption. They are called queueing with set up.)

When service is completed, the next unit in the queue to be serviced is chosen and under the usual assumptions, that unit immediately goes into service. The rule used to choose the next unit to serve from among all those in the queue is called the <u>queue discipline</u>. The most common assumption in queueing theory is that the first unit in the waiting line is the first customer to be served. Such a discipline is called a <u>first in-first out discipline</u>. For other applications other disciplines have been considered. (In military studies the next unit to be served may be the one posing the most immediate threat. In a computer center, jobs may be assigned to classes of importance - called priority classes - and the next job chosen is from among those with the highest priority. In production systems the next job to manufacture may be the one whose promised delivery date - the so called "due date" - is closest.)

There are four basic processes created by the interaction of service times, interarrival times and queue discipline that are of particular importance though others have been studied. The length of the waiting line, called the queue length process, $\{N(t): t \geq 0\}$, is one of the basic processes of concern. The length of time from entry to exit of the queueing system, $\{Q_n: n = 1,2,\cdots\}$, called the sojourn time process, is another of the basic processes. The time from first entrance of a unit to an idle server until first entrance of the next unit to an idle server, $\{B_n: n = 1,2,\cdots\}$, is called the busy cycle. The time between consecutive exits from the server, $\{X_n: n = 1,2,\cdots\}$, is called the departure process.

Somewhat more formally we can define the problems as follows:

(1) let $0 \leq T_0 < T_1 < T_2 \cdots$ be a sequence of real random variables representing the times at which an arrival occurs to a queueing system. Let $A_n = T_n - T_{n-1}$, $n = 1,2,\cdots$. Assume $\underset{\sim}{A} = \{A_n\}$ is a sequence of i.i.d. random variables (a renewal process). Call $\underset{\sim}{A}$ the arrival process to the queue;

(2) let S_n be a non-negative real random variable. Let $S = \{S_n: n = 1,2,\cdots\}$ be a sequence of i.i.d. random variables. Call $\underset{\sim}{S}$ the service time process for the queue. We assume $\underset{\sim}{A}$ and $\underset{\sim}{S}$ are independent processes;

(3) let $1_{[0,t]}(T_n)$ be an indicator random variable taking values 0 (or 1) if T_n does not (or does) occur in $[0,t]$;

(4) define $N_A(t) = \sum_{n=0}^{\infty} 1_{[0,t]}(T_n)$. $\{N_A(t): t \geq 0\}$ is called the arrival counting process.

The four processes of basic concern to queueing theory can be defined formally in terms of the $\underset{\sim}{A}$ and $\underset{\sim}{S}$ process. Prabhu [1965] provides a formal definition of the queue length process. We will not reproduce that here but rather we assume that everyone has an intuitive understanding of such a process. The sojourn time process is closely related to the waiting time process for first in-first out disciplines which can be formally stated as

$$W_n = W_{n-1} + S_{n-1} - A_n, \quad \text{if } W_{n-1} + S_{n-1} - A_n > 0,$$
$$= 0 \quad , \quad \text{otherwise.}$$

Then $Q_n = W_n + S_n$ defines the sojourn time of the nth arriving customer. The busy cycle process is the sequence of passage times already discussed. If we define X_n as the time between the nth and $(n-1)^{st}$ service completions then the departure process, $\{X_n : n = 1,2,\cdots\}$, can be more formally given as

$$X_n = \begin{cases} S_n, & \text{if the } (n-1)^{st} \text{ unit leaves someone in the queue,} \\ I_n + S_n, & \text{if the } (n-1)^{st} \text{ unit leaves an empty queue.} \end{cases}$$

Here I_n is the foreward recurrence time in the arrival process measured from the time of departure of the $(n-1)^{st}$ unit. I_n is called the idle time of the server.

1.3 Special Cases. The major thrust of queueing theory for much of its life has been to compute formulas for the above measures of effectiveness with special assumptions on the arrival process, service process, queue discipline or system capacity. For future reference, we will record some results for one special case here. The following results as well as many of the special cases are discussed in most standard texts on the subjects. (For example, see Kleinrock [1975; 1976].)

If we assume that $N_A(t)$ is a Poisson process (with parameter λ as is usual) and S_n is exponentially distributed (parameter μ as is usual), the service center has 1 server, the queue discipline is first in-first out and the system's capacity is unlimited then the system is called an M/M/1 queueing system. A considerable amount is known about the above four processes (Kleinrock [1975]). For example:

(1) $\{N(t)\}$ is a Markov process with state space $(0,1,2,\cdots)$;

(2) $\Pr[N(t) = k]$ is known (but we'll not reproduce the results here. Essentially the probability is given by an infinite sum of Bessel functions of

type - I, i.e., Bessel functions of the second kind with imaginary argument.);

(3) $\lim_{t\to\infty} \Pr[N(t) = k] = (1-\rho)\rho^k$, $k = 0,1,2,\cdots$ for $\rho = \lambda/\mu$ (called the traffic intensity). Such a limiting distribution exists if and only if $\rho < 1$, otherwise the limit is identically 0 for all finite k. These limiting probabilities, when they exist, are called the steady state probabilities;

(4) $\{W_n\}$ is a random walk on the non-negative reals with a delaying barrier at 0. (See Feller [1963] or Borovkov [1976].);

(5) $\Pr[W_n \leq t]$ is known in principle (through the theory of random walks);

(6) $\lim_{n\to\infty} \Pr[Q_n \leq t] = 1-e^{-(\mu-\lambda)t}$, $t \geq 0$. This limit exists if and only if $\rho < 1$, otherwise the limit is 0 for all finite $t \geq 0$;

(7) the busy cycle process is a sequence of i.i.d. random variables whose distribution is known (Cox and Smith [1961], for example);

(8) the departure process $\{X_n: n = 0,1,2,\cdots\}$ is, in principle, known for all n from (2);

(9) in the steady state (i.e., $t \to \infty$), when $\rho < 1$, $\{X_n\}$ is a sequence of i.i.d. random variables that are exponentially distributed with parameter λ. That is, the departure counting process is a Poisson process with the same parameter as the arrival process. (See Burke [1956] or Disney, et al. [1973].)

Results (1), (3), and (9) are useful to our future discussion.

2. Jackson Networks

2.0 Introduction. It was recognized early in queueing theory that a theory of single server systems was not adequate. Indeed, at least by the 1930's (and before if one is not too picky) research in queueing had begun to explore systems of queues or what we shall call networks of queue. Erlang about 1910 had studied systems where many identical servers all handled the same arrival process (the M/M/c model) in telephone systems. A good discussion of these and multiple server models is given in Syski [1960], especially chapters 5 and 6 as well as chapters 7 to 10.

2.1 Jackson Networks. A general model in this historic mold that has served almost as the definition of a queueing network was studied in a paper by J. R. Jackson [1957]. These models and related problems have come to be called "Jackson networks." His 1957 model assumed:

(1) there are $1 \leq M < \infty$ service centers (Jackson did allow servicing center j to be comprised of more than 1 server. However, to keep the problem simple we assume throughout that each servicing center has only one server.);

(2) these M servers are arbitrarily connected by arcs over which units travel instantaneously fast;

(3) to establish how units proceed through the network one defines p_{ij} to be the probability that a unit completing service at service center i proceeds to service center j for its next service. Then for each i, $1 - \sum_{j=1}^{M} p_{ij}$ is the probability that a unit exiting node i leaves the system. The matrix $\underset{\sim}{P}$ whose elements are p_{ij} is called the switching matrix for the network. Note that these assumptions about switching imply that a unit leaving service center i chooses the next service center to visit without regard to any other condition. We call this switching behavior a multinomial switch;

(4) service times, S_{ni}, $n = 1,2,\cdots$, at center i are i.i.d. random variables. These times are each exponentially distributed random variables with parameter μ_i. The sequences $\{S_{ni}\}_{i=1}^{M}$ are independent sequences;

(5) there may be many arrival processes but each one is a Poisson process (with parameter λ_i if the arrival process is to service center i). These arrival processes are independent of each other and of the network;

(6) all queue disciplines are first in-first out;

(7) all queue capacities are unlimited.

2.2 The Queue Length Process. The results that Jackson obtained were rather surprising. Define $\underset{\sim}{N}(t)$ to be a vector whose j^{th} element, $N_j(t)$, $j = 1,2,\cdots,M$, is the queue length at the jth service center at t.

Jackson's major results were as follows:

(1) $\{\underset{\sim}{N}(t): t \geq 0\}$ is a vector-valued Markov process;

(2) $\lim_{t\to\infty} \Pr[N_1(t) = k_1, N_2(t) = k_2 \cdots N_M(t) = k_M]$

$= \lim_{t\to\infty} \Pr[N_1(t) = k_1]\Pr[N_2(t) = k_2]\cdots\Pr[N_M(t) = k_M].$

That is, the queue length processes in these Jackson networks are asymptotically ($t\to\infty$), mutually independent;

(3) $\lim_{t\to\infty} \Pr[N_j(t) = k_j] = (1-b_j)b_j^{k_j}$, $j = 1,2,\cdots$ if and only if $b_j < 1$, otherwise this limit is 0. Here $b_j = a_j/\mu_j$ and a_j satisfies the so-called traffic equation

$$\underset{\sim}{a} = \underset{\sim}{\lambda} + \underset{\sim}{P}\underset{\sim}{a}.$$

$\underset{\sim}{a}$ is the M-vector of a_j, $\underset{\sim}{\lambda}$ is the M-vector of λ_i, and $\underset{\sim}{P}$ is the switching matrix. We assume throughout that the traffic equation has a positive solution (or alternatively the maximum eigenvalue of $\underset{\sim}{P}$ is less than 1). In elemental form, this equation says simply that a_j is the rate of arrivals to service system j. This arrival rate is simply the sum of exogeneous arrivals to system j, (λ_j), plus the sum of all arrivals to j from inside of the system, $(\underset{\sim}{P}\underset{\sim}{a})_j$.

It is instructive to spend a moment looking at what Jackson said and did not say. Many papers have been written on Jackson networks that misinterpret and misuse his results. The major confusion occurs around the third result. Remember that in the example of section 1.3 we showed that the queue length process was asymptotically ($t\to\infty$) geometrically distributed with a parameter $\rho = \lambda/\mu < 1$. If $\rho \geq 1$ the limit is zero. Also recall that λ was the parameter associated with the Poisson arrival process and μ was the parameter associated with the exponential service times.

Now result (3) above is also asymptotically geometrically distributed with a parameter a_j that is acting as an arrival rate and a parameter μ_j acting (is) as a servicing rate. This result led Jackson to state "This theorem (our results (2) and (3) above) says, in essence, that at least so far as steady states are concerned ($t \to \infty$), this system with which we are concerned behaves *as if* its departments (our centers) were independent, elementary (i.e., M/M/1) systems ...". So far, so good. The *as if* emphasis is Jackson's and as it stands is innocuous enough. Jackson makes one more statement about his results that may have caused an enormous amount of confusion. He says "This conclusion is far from surprising in view of recent papers by E. J. Burke (that should be P. J. Burke) and E. Reich." It appears from his references that Jackson is here alluding to the result (9) of section 1.3 which is sometimes called "Burke's Theorem".

Confusion in queueing network literature has occurred by taking the "as if" of Jackson too seriously. Researchers seem to have taken this to mean that the queue length processes are independent and are created by M/M/1 subsystems. As a consequence there are papers in the queueing literature that study queueing networks one node at a time (the independence assumption) as M/M/1 queues (the assumption of Burke's theorem). There are other papers that study sojourn times in networks as sums of independent, exponential random variables. (See section 1.3, result (6).) Unfortunately, many authors have shown that the flow of units within the network are, except for special network configurations such as trees, not Poisson processes and in fact, are not even renewal processes. Thus each service center in isolation is not only not an M/M/1 queueing system, it is not even a queueing system with a renewal arrival process. That is, for a time there was serious confusion in the literature as to what *as if* really meant. The problem is fairly well understood now. Section 5 and its references discuss the topic more thoroughly.

2.3 Waiting Times. Waiting times in Jackson networks is a rather unexplored area. Attention has recently turned toward that area but research here has only started. There are few special cases that have been studied in detail principally in what are called tandem queues.

We can define a tandem, Jackson network as one having the structure of section 2.1 with the added features:

(1) there is only one arrival process and it occurs to the first server (In terms of the section 2.1 scheme $\lambda_1 = \lambda$, $\lambda_j = 0$, $j \neq 1$.);

(2) $p_{ij} = 0$ unless $j = i + 1$ in which case $p_{ij} = 1$, $j = 1,2,\cdots,M$. (See figure 1.)

FIGURE 1

A TANDEM JACKSON NETWORK

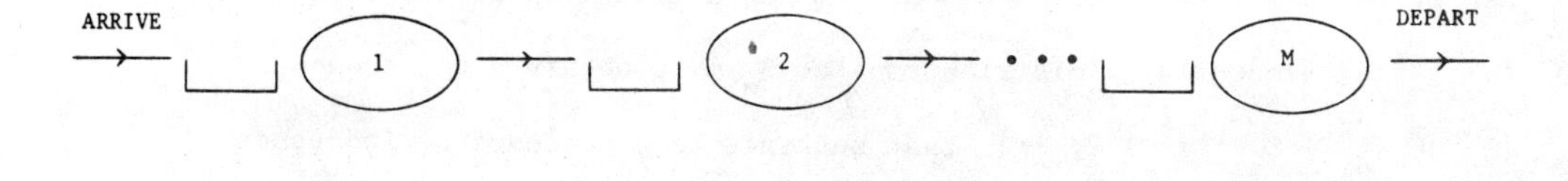

J — SERVICE CENTER

UNLIMITED CAPACITY QUEUE

DIRECTION OF FLOW

Since the only results, with which we are familiar, are concerned with first in-first out queue disciplines, we assume this discipline throughout. We need a bit more symbolism here. So define:

W_{ni} = the total time unit n spends waiting in service center i. Call this the <u>waiting time</u> of unit n in that service center;

Q_{ni} = the time spent by unit n in service center i. Call this the <u>sojourn time</u> of unit n in center i. $W_{ni} + S_{ni} = Q_{ni}$, $i = 1,2,\cdots,M$, $n = 1,2,\cdots$.

The first result is from Reich [1957] which shows: for fixed n, in the steady state, $\{Q_{ni}\}$ is a sequence of i.i.d. random variables, each of which is exponentially distributed. Thus, if Q_n is the sojourn time in the system for unit n, this result says simply that this sojourn time, in the limit, is the sum of independent, random variables

$$Q_n = Q_{n1} + Q_{n2} + \cdots + Q_{nM}.$$

The Q_{ni} may depend on μ_i. Nonetheless the distribution of Q_n can be obtained from simple convolution operations.

The next result is surprising. It comes from Burke [1964]. For the tandem queueing system he shows: the sequence $\{W_{ni}\}$ for n fixed and $i = 1,2,\cdots,M$ is, in the limit, a sequence of dependent random variables. That is, sojourn times at successive service centers in the network are, in the steady state, independent but waiting times are not. It is curious that the sequence $\{W_{ni}: i = 1,2,\cdots,M\}$ for fixed n is a sequence of dependent random variables. But if to each term in this sequence we add a random variable, independent of $\{W_{ni}\}$ and exponentially distributed (the S^{*}_{ni}) to obtain a new sequence, $\{Q_{ni} = W_{ni} + S_{ni}: i = 1,2,\cdots\}$, that sequence is a sequence of independent random variables for each n.

These waiting time problems exhibit another nasty property that has been proven in two distinct cases. Burke [1969] proved the first result. To expose it we must assume that a service center can have more than one server. It is usual to assume that all service times at a given center are mutually independent and all are exponentially distributed with parameter μ_j that depends only on the center (not on the server in the center - the servers are "identical"). With this set up, Burke considers a three center tandem queue with the first center having just one server, the second center having any finite number of servers, (> 1), and the third center having just one server. Otherwise, the problem is that of Reich discussed above. Burke is then able to show that for each n:

Q_{n1} and Q_{n2} are independent random variables;

Q_{n2} and Q_{n3} are independent random variables;

Q_{n1} and Q_{n3} are not independent random variables.

These results are surprising. In this network one does not have the Reich mutual independence. In fact, one does not even have pairwise independence.

Simon and Foley [1979] have found a similar result to that of Burke in a different network. In their three service center network, there is one server at each center. $\lambda_1 = \lambda$, $\lambda_j = 0$, $j = 2,3,$. The new idea is that the network is not a tandem queueing network. Rather one has $p_{12} = p$, $p_{13} = 1 - p$,

$p_{23} = 1$, $p_{3j} = 0$, $j = 1,2,3$ in the structure of section 2.2. (See figure 2.) Then they are able to show that:

Q_{n1} and Q_{n2} are independent random variables;

Q_{n2} and Q_{n3} are independent random variables;

Q_{n1} and Q_{n3} are not independent random variables.

FIGURE 2

THE SIMON-FOLEY NETWORK

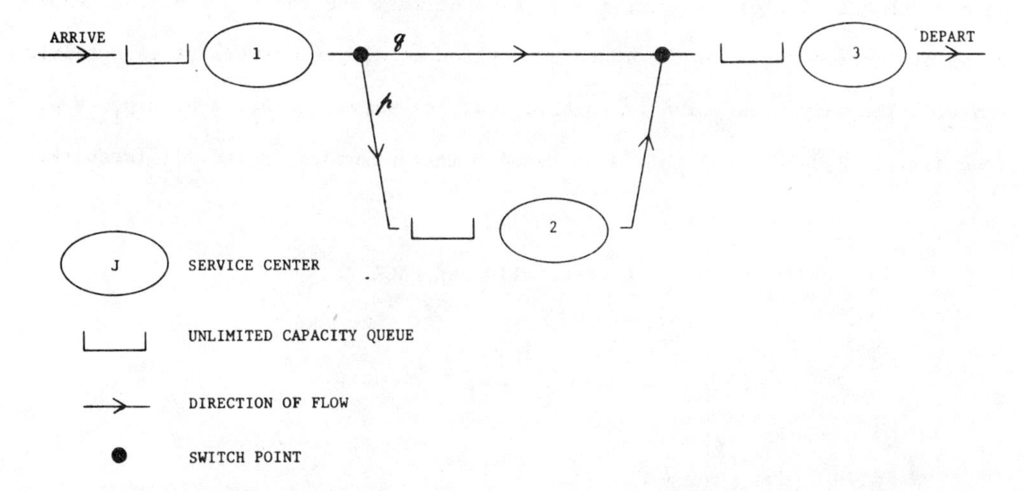

One conjectures that what is happening here, in both the Burke problem and the Simon - Foley problem, is that a unit in queue 2 (in those examples) may be by-passed by units in server 1 that arrive to the system after unit n. In both cases, then, the queue length and hence the sojourn time at server 3 will depend on how many such units by-pass unit n, which in turn depends on how long unit n was in server 1.

Following this line of thought, the following conjecture appears reasonable. If there is only one path connecting any two single service centers in the network then one can use the result of Reich [1957] to determine the total sojourn time of a given unit through a Jackson network. If, on the other hand, there are multiple paths connecting single server service centers then the sojourn times of unit n at the service center at the start of these multiple paths and that at the termination of the multiple paths are dependent. Furthermore, if there can be multiple servers at any service center then the sojourn times at service centers feeding this multiple server service center and those being fed by it are dependent unless the multiple server center is first or last in the sequence.

Unfortunately, we do not know the nature of these dependencies. Neither do we know the sojourn times through either the Burke or Simon and Foley networks. This is an area in need of considerably more study.

There is yet one more problem with sojourn times in these Jackson networks. In perhaps the simplest non-trivial Jackson network one takes all of the assumptions of section 2.1 with the added assumption of single servers at each service center. The only alteration is to have just one service station with $p_{11} = p$. (See figure 3.) We call this a queue with instantaneous, Bernoulli feedback.

FIGURE 3

A QUEUE WITH FEEDBACK

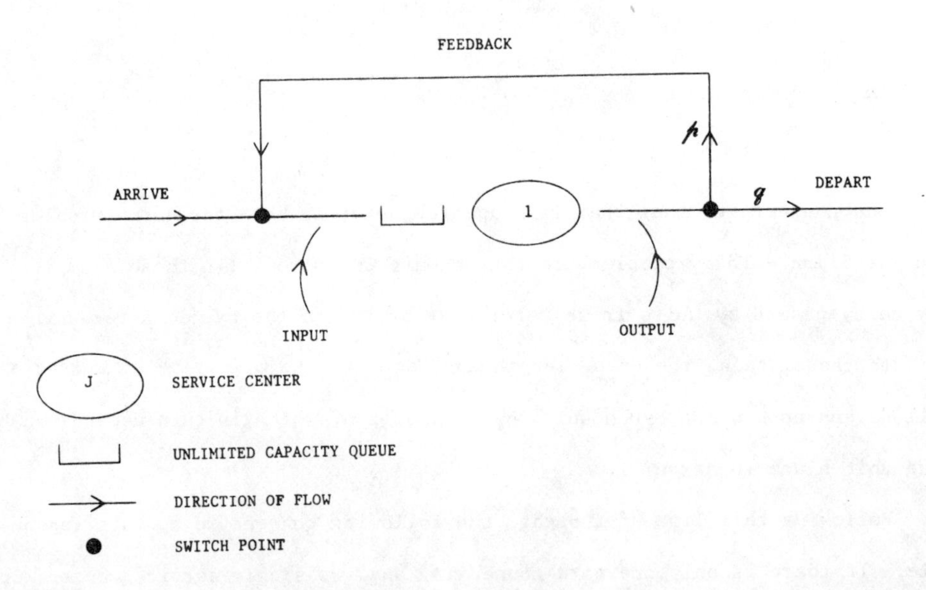

The definition of sojourn time needs a modification. So let: Q_n^i = the sojourn time of unit n the ith time that unit passes through the server. Then, Q_n = the sojourn time of unit n.

$$Q_n = Q_n^1 + Q_n^2 + \cdots + Q_n^k,$$

where k is a geometrically distributed random variable.

It has been shown by Takacs [1963] and Disney [1978] that for each fixed n, $\{Q_n^i, i = 1,2,\cdots\}$, is a Markov renewal process. Takacs and Disney each solve the problem for slightly more general cases than Jackson networks using rather different methods. Once again, as in the Burke or Simon and Foley examples, the lack of independence seems to be coming from what we loosely called by-passing.

2.4 Summary of Results. In summary, unlike the Jackson queue length process results that have been obtained and are elegant in appearance, waiting time and sojourn time properties are largely unknown. For tandem queues studied by Reich the problem can be considered solved. For general, single server, Jackson networks that do not have multiple paths between any two service centers in the network, it appears that the Reich results can be used to obtain results easily. For more general networks of Jackson type, the sojourn time problem is largely unsolved. Because of its importance to many areas it is one of the most pressing problems in queueing network theory.

3. Busy Periods and Departure Processes

3.0 Introduction. The queue length process and sojourn times have been the major topics of interest to queueing theory since its beginning. While of some importance to systems analysis, the busy period and to a lesser extent (until rather recently) the departure process from single service queues have been of less importance. The same result is true for Jackson queueing networks as we shall see below.

3.1 The Busy Period. One can define two related processes of interest. One is called the busy period. The other is called the busy cycle. The busy period is the time from entrance of a unit to an empty queue to exit of a unit that leaves the system empty. The busy cycle is the return time from entrance of a unit to an empty queue to the next time of entrance of a unit to an empty queue. If the queue is ergodic such points occur infinitely often. If the queue is not ergodic such points, except possibly the first one, may not exist. Almost all studies of the busy cycle and busy period with which we are familiar, assume these times occur infinitely often. (See Cox and Smith [1961].)

In single server queueing theory in which $\{A_n\}$ is a sequence of i.i.d. random variables, $\{S_n\}$ is also a sequence of i.i.d. random variables and $\{A_n\}$ and $\{S_n\}$ are independent sequences, the sequence of busy cycles, $\{B_n : n = 1,2,\cdots\}$, is a sequence of i.i.d. random variables if the origin of the time scale is taken as a point of arrival to an empty queue. The busy period sequence is not a sequence of i.i.d. random variables. General distributional results are known for the cases in which the two processes are sequences of i.i.d. random variables. (For example, see Kleinrock [1975].)

Unfortunately, the related problems in queueing networks are in a much less well developed form. To the best of our knowledge there are no known results for busy periods or busy cycles for Jackson networks. The area could stand some study.

Stating the problem that needs to be considered is rather easy. Recalling from section 2.2 (item 1) that $\{N(t)\}$ is a vector valued Markov process, the busy cycle process is then simply the time from entrance of an arriving unit to an empty network to the next time of entrance of an arriving unit to an empty network. In the busy period case the problem is to determine the time from entrance of an arriving unit to an empty network to the time at which a departing unit leaves behind an empty network. We surmise that such points occur infinitely often for ergodic networks but as pointed out above we know of no results for the busy period or busy cycle of a Jackson queueing network.

3.2 Departure Processes. There are two problems here that lead to interesting questions related to our discussion of the queue length process and waiting time process. One problem is concerned with the nature of the departure processes from the Jackson network. The other problem is concerned with the nature of the departure processes from single centers in the network. (Daley [1976] is a nice review of these departure processes.) To simplify our discussion we will assume the network is irreducible in the sense that the switching matrix $\underset{\sim}{P}$ is an irreducible matrix. More general cases have been studied. The interested reader is referred to Melamed [1979] and its references for these other cases.

When $\rho_i < 1$, for $i = 1,2,\cdots,M$, every entering unit eventually leaves the system. Then from Melamed [1979] we have:

(1) in Jackson networks with single server service centers, if j is a node from which departures from the network occur then the departure process from the network at this center is a Poisson process with parameter a_i;

(2) the collection of Poisson processes of departures from the network are mutually independent;

The first of these results is quite in keeping with the Burke theorem (section 1.3, item (9)). The second result is, at first glance, rather surprising. One would expect that the network itself imposed some dependencies on the departing processes.

When one turns to consider departure processes from individual service centers in the network things became a bit more complicated. For the time being we will continue the discussion within the framework of Jackson networks. Furthermore, we must distinguish two cases. If $p_{ii}^{(n)} > 0$ is defined as the n step transition probability from i to i in the switching process, then there is some path leading from service center i back to that service center. In this case we will say that service center i has <u>feedback</u>. Otherwise we will say service center i does not have feedback. The latter case we can dispose of quickly.

If service center i does not have feedback, the departure process is a Poisson process with parameter a_i.

The service center with feedback requires distinguishing two processes. (See figure 3 for a picture of the terms used here.) In one process, units leaving service center i will eventually return to i. In the other process, units leaving service center i will never return to i. Call the former process, the <u>feedback stream</u> and the latter the <u>departure stream</u>. (See figure

3.) Then again from Melamed [1979] we have:

(1) the departure stream is a Poisson process;

(2) the feedback stream is not a Poisson process and in fact is not a sequence of i.i.d. random variables.

We can flesh out item (2) a bit more in the case $p_{ii} > 0$, that is, feedback occurs in one step - the so-called instantaneous feedback case. In that case the output process (figure 3) is a Markov renewal process whose transition functions are known. Then it has been shown by Disney, et al. [1980] that the output process is never a Poisson process nor is it a renewal process. However, the departure process is a Poisson process (parameter λ). The feedback process is not a renewal process. There is reason to believe that the departure process and the feedback process are not independent random processes but we know of no proof either way here.

These Melamed and Disney et al. results raise some interesting anomalies In these Jackson networks, as was noted in section 2.2 (item 3), queue length at individual service centers act as if they were independent, M/M/1 queues. Yet if the network has feedback loops, the flow on these loops is not a Poisson process nor even a sequence of i.i.d. random variables. It is this property (i.e., the distinction between properties of the network and properties of the individual service centers in the networks) that seems to have created confusion in some applications of the Jackson network results both in the study of the queue length process and that of the waiting time process.

4. Extensions to the Jackson Network Theory of 1957

4.0 Introduction. Following his 1957 paper, Jackson next published a paper in 1963 in which he followed the basic ideas of the earlier paper. The new ideas were to allow the arrival processes to the network to be birth processes whose parameters could depend on the total number of units in the network. Similarly, the service time processes were death processes with parameters depending on the number of units at a given service center. In this way the queue length process $\{\underset{\sim}{N}(t)\}$ becomes a vector valued birth-death process.

4.1 Queue Length Processes in the Vector-Valued Birth-Death Process. We will not reproduce the exact form of Jackson's 1963 results. They would require introducing a large amount of new symbolism. Currently available work which we shall discuss later includes these results. However, it is important to summarize the findings of Jackson (at a cost of imprecision) because they have led many others to explorations in queueing networks in an attempt to generalize the concept of a Jackson network.

In his 1963 paper, Jackson finds the steady state probability vector for $\{\underset{\sim}{N}(t)\}$. Much as in the 1957 paper the elements of this vector have a geometric-like form (called a product form in much of the computing literature). That is, the elements are of the form

$$C\, b_1^{k_j}\, b_2^{k_2}\, b_3^{k_3} \cdots b_M^{k_M}, \quad k_j \geq 0.$$

In the 1957 paper the constant C had the form

$$(1-b_1)(1-b_2)\cdots(1-b_M)$$

which leads directly to the results, observations, and comments found in section 2. In the 1963 case, however, C is not found to be of this form (except in special cases such as those in the 1957 paper). As a consequence, one cannot infer that these networks are composed of independent single service centers nor that these service centers act as if they were simple queues.

Since these results appeared, considerable effort has been expended trying to determine approximations and easy ways to compute C. Other research effort has been expended on exploring networks that might have the "product form" of solution. There are no up to date summaries of the large amount of work. The Kleinrock [1975; 1976] books are basic. The papers of Kelly [1976; 1978] and Schassberger [1977; 1978] trace some of the work.

4.2 A Generalization. In a 1976 paper, Kelly significantly generalizes the concept of Jackson network and provides important extensions to the concept of "product forms" of solutions. In Kelly [1979], he provides further insights and significantly broader applications of his study. We will follow his 1976 publication.

Suppose that there can be I types of units entering the network. Units of type $i \in I$ enter the network as a Poisson process with rate $\nu(i)$ and pass through the servers according to the path

$$r(i,1)r(i,2)\cdots r(i,S(i))$$

before leaving the system. Thus at stage s $(s = 1,2,\cdots S(i))$ of its route, the unit is at queue $r(i,s)$.

Within each queue, the units are ordered so that there are units in positions $1,2,\cdots,n_j$. n_j is the total number of units in queue j.

Each unit requires a random amount of service. This service time is an exponentially distributed random variable with mean 1.

There is a single server who supplies a total service effort at rate $\phi_j(n_j)$ in such a way that $\gamma_j(\ell,n_j)$ of this effort is directed to the unit in position ℓ. When this unit leaves the jth service system, all units behind it move up a space.

When a unit arrives at service center j it moves immediately into location ℓ with probability $\delta_j(\ell, n_j + 1)$. Units formerly occupying spaces ℓ, $\ell + 1,\cdots,n_j$ are moved to spaces $\ell + 1$, $\ell + 2,\cdots,n_j + 1$.

Such a structure is rather general for queueing network behavior. The queue discipline of the earlier Jackson networks has been considerably generalized. Arrival processes are still Poisson but note the arrival rate may depend on the "type" of the unit. "Type" may be associated with the path taken by the arrival simply by associating an arrival "type" with the path that arrival will follow. Service times here are also more general.

Another important aspect of the Kelly paper is its method of determining the limiting probability vector for the queue length random process. Whereas the earlier Jackson network papers of 1957 and 1963 proceeded from the structure of the problem to set up the usual limit form of the Kolmogorov equations of the Markov process $\{\underset{\sim}{N}(t)\}$, Kelly prefers to work with relations embodied in a reversed process. It appears that when such an approach can be made to work, detailed calculations are obviated. The Kelly book [1979] expands on this point and considers these methods for a wide variety of problems in multidimensional Markov processes, including the queueing networks of his 1976 paper.

Kelly starts with a stable, conservative, regular Markov process. (All of the networks discussed so far have these properties.) Then it is well known from the theory of such processes that if i, j are vector valued states in these networks, then a solution to the equations

$$\Pi \mathbb{P} = 0$$

with $\Pi \geq 0$ and $\Pi \underset{\sim}{1} = 1$ is unique and

$$\Pi(j) = \lim_{t\to\infty} \Pr[\underset{\sim}{N}(t) = j].$$

The elements of Π are the limiting probabilities - the so-called "steady state" probabilities - of the Markov process. The elements of $\mathbb{P}$ are the transition rates of the process (Kleinrock [1975]). $\mathbb{P}$ is called an infinintesimal generator.

Now it is known that if $\{\underset{\sim}{N}(t)\}$ is a Markov process in equilibrium (e.g., the initial vector for the process is Π) there then exists another process - called the reversed process, $\{N(-t)\}$, that is a Markov process in equilibrium. The reversed process has the same limiting probability vector, Π, but its infinitesimal generator may not be that of $\{N(t)\}$. (If the two processes have the same infinitesimal generator then $\{N(t)\}$ is said to be reversible.) In general if $q(i,j)$ is the (i,j) element of the infinitesimal generator of $\{N(t)\}$ and $q'(i,j)$ that of the infinitesimal generator of $\{N(-t)\}$ and if $q(i)$, $q'(i)$ are the corresponding diagonal elements of the infinitesimal generators of $\{N(t)\}$, and $\{N(-t)\}$ respectively, then we have

$$\Pi(i)q(i,j) = \Pi(j)q'(j,i) \tag{4.2.1}$$

and

$$q(i) = q'(j).$$

(If the process is reversible these equations are called the equations of detailed balance.)

The extremely useful result is that for the network set up by Kelly, one can find $q(i,j)$ and $q'(i,j)$ rather easily. This, along with (4.2.1) and the uniqueness of Π as a probability vector then allows one to determine Π.

For his network, Kelly defines a two-tuple $c_j(\ell) = ((t_j(\ell), s_j(\ell))$ where $t_j(\ell)$ denotes the "type" of unit in position ℓ in queue j and $s_j(\ell)$, as previously defined, represents the position along its path reached by this ℓ^{th} unit in queue j. It is shown that for the vector

$$c_j = (c_j(1), c_j(2)\cdots c_j(n_j)),$$

$$\underset{\sim}{C} = (c_1, c_2, \cdots, c_M)$$

is an irreducible Markov process on a countable state space. The infinitesimal generator for this process and its reversed process are found. Then using the results on reversed processes stated above, Kelly shows that his network has the product form of solution which is determined up to a normalizing constant.

4.3 Another Generalization. Kelly generalizes his model one more step and in the process provides the basis for taking a major step out of the restrictive assumption of Poisson processes and exponential service times. Unfortunately, this step has a price to pay.

The first step is to generalize the service time assumption of section 4.2. Now instead of service times being exponentially distributed random variables, it is assumed that when at queue j, the unit that is at stage s of its path requires an amount of service that is the sum of $Z(j,s)$ independent, identically distributed, exponential random variables (rate $d(j,s)$). That is, service times are now gamma distributed random variables with parameters $Z(j,s)$ and $d(j,s)$.

The other assumptions of section 4.2 are retained except that it is required that

$$\delta_j(\ell, n_j + 1) = \gamma_j(\ell, j_j + 1). \qquad (4.3.1)$$

For this problem the state space of the Markov process of interest must now become a three-tuple. A state now is defined by $t_j(\ell)$, $s_j(\ell)$, as in section 4.2, and $x_j(\ell)$ which denotes the phase of service currently occupied by the unit. Then, on this three-tuple space one can define a Markov process whose states are the three-tuples. This process is irreducible and the state space is countable. The process has a reversed process; its infinitesimal generator and that of the reversed process can be found. As before the properties of reversing are used and it is shown that the process has a product form of solution.

4.4 A Major Generalization. Were the Kelly results to stop here they would make interesting, perhaps useful, contributions to the theory of Jackson queueing networks. The restriction of Poisson arrival processes and either exponentially distributed or gamma distributed service times would preclude a wide spread use of the results (although much of the computer systems analysis literature finds these conditions to be reasonable for many computer studies). But more is available.

Under the conditions of the model of section 4.3, Kelly conjectures that his results can be extended to include G-type service times. (See section 1.2.) The conjecture is based on a result of Whitt [1974] which shows that finite mixtures of gamma distributions are dense in the set of arbitrary non-negative distributions. Though Kelly does not prove that his model of section 4.3 extends to G-type servers, it is proven with the requisite care by Barbour [1976].

Finally, Kelly drops the Poisson arrival process assumption that has run through his work. Commenting on his models (described in our section 4.2 and 4.3) he notes that one can allow these arrival processes to be birth processes whose parameter depends on the total number of units in the system. Recall that this step was made in the Jackson [1963] paper.

4.5 Comments. The Kelly work probably represents the state-of-the-art in the study of queueing networks originally arising out of the papers of Jackson.

Work continues on these problems and the Kelly work will probably be mined for quite a while. If one could drop condition (4.3.1) from the Kelly model and still be able to compute an equilibrium solution, one would have a major contribution to the theory of queueing networks.

Concerning waiting time, busy period, and busy cycle analysis we know of no results presently available. Concerning departure processes, Kelly presents some results on the departures from the network. These processes are independent, Poisson processes for the models studied in our sections 4.2 and 4.3.

It has been noted by many authors (for example, see Kelly [1976]) that since the Jackson limiting probability vector depends on the assumptions of the arrival process and service time process only through the expected values of these processes, similar results may well hold for more general arrival and service time processes. That is, results such as those obtained may be independent of the distributional assumptions of the model. Schassberger [1978] explores this topic in more detail and provides a bibliography for further reading. He calls this independence property, insensitivity.

5. Flow Processes

5.0 Introduction. As soon as one moves very far from the Jackson assumptions given in section 2, analysis of queueing networks become difficult, at best. Yet in many applications these assumptions are inappropriate. One would like to extend the Jackson model.

Relaxing the assumptions presents enormous difficulties. In many cases, of course, one can augment what is meant by a "state" of the process to retrieve the useful Markov property inherent in the Jackson and Kelly models. But, for example, if one were to retain all of the Jackson assumptions of 1957 except that the service times were allowed to be non-exponentially distributed (and thereby, the "forgetfulness" property is lost as is the Markov property for $\{\underset{\sim}{N}(t)\}$), one requires a 2M-tuple to retrieve the Markov property for $\{\underset{\sim}{N}(t)\}$. Furthermore, this 2M-tuple would have the Markov property on a cross product space. The space of M-tuples representing the queue length at each service center is the non-negative lattice points in E^M. This space would have a Cartesian product with the space of M-tuples representing expired service times at each service center. Elements of these M-vectors take values in the non-negative E^M space. This cross product space is not an appealing one to work with even for $M = 1$.

5.1 Another Approach to Queueing Networks. To extend the basic models discussed so far it seems that one must adopt a new approach to the entire problem. One such approach is to decompose the network into smaller subnetworks that, hopefully, are easier to study. Ultimately, of course, one must recompose the network.

If one steps back and looks at what the network is doing, it seems as though these networks can be described basically as a collection of three operations being performed on a set of arrival processes. We call these three operations random deletion, random stretching and recomposition. They are discussed in section 5.2. Queue lengths, busy periods, waiting times are consequences of these operations and in principle should be approachable if the consequences of the operations are known.

The basic problem in studying these networks as smaller subnetworks, is that one does not know the stochastic properties of the arrival process to any node in the network, except possibly those nodes that serve only as entrances for exogenous arrivals or for nodes in very special networks (e.g., trees). In most cases the operations of the network itself serve to transform one stochastic process (e.g., the arrival process) into another stochastic process with rather different properties. For example, the departure process from an M/G/1/N queue is, except in four cases, never a renewal process. The arrival process which is a Poisson process, and hence a renewal process, is transformed by the queueing system into a non-renewal process.

5.2 Operations on Stochastic Point Processes. In an abstracted form one can view a queueing network as a collection of three operations performed, in some sequence (defined by the network), on one or more marked point processes (the arrival process). These operations are known in point process theory as: random deletions, random stretching, and superposition.

To keep the discussion focused, consider the Jackson network. In that case, the multinomial switches (assumption 3, section 2) essentially operate on a marked point process (the Poisson arrivals or departures from a given service center, i) by deleting those points (and their marks) from the point process that will be arrivals to service center j. The undeleted points constitute departures from the network or arrivals to some service center, not j. Thus, the switching matrix in the Jackson network is, essentially, a structure for randomly deleting points from a marked point process. The related switching structure in the Kelly network is a more complicated set of rules for effecting deletions.

Furthermore, in a queueing network one can suppose that the arrival process to a service center is a marked point process. Then, what the service center itself does is to randomly delay each point in this process to form a new marked point process called the departure process. The amount of delay is the random amount of time that an arrival spends in the service center.

Finally, one can view the arrival process to a service center as a generalization of what in stochastic point process theory is called superposition. There are two ways to view the problem. For our purposes one has several sets of points representing the epochs of arrival to the queue. If now one forms the union of these sets and then orders these points (increasing order), the resulting ordered set is called the superposed set. The sets thus superposed can be called the constituent sets. Superposition is then concerned with the study of the properties of the superposed set. In nearly all studies of superposition (there are a few exceptions), the constituent sets are assumed to be independent marked point processes. The elements within the constituent sets need not be independent, however.

In the studies such as Jackson [1963], some of the constituent processes are neither point processes, as that term is usually used, nor are they independent. Thus, for a more general class of problem, we use the word recomposition rather than superposition. So the operations we want in the study of queueing networks are operations that recompose collections of marked point processes.

Thus, one approach to the study of queueing networks in which $\{\underset{\sim}{N}(t)\}$ the vector of queue lengths is not a Markov process (e.g., "non-Jackson" networks) is to look at flows of units in the network as marked point processes and to look at the network as a collection of the operations of random deletions, random stretching and recomposition. In this way attention is turned, temporarily from the queueing properties of queue length, busy periods or cycles, and waiting times and given to the study of "random flows" in queueing networks. In this way one can generalize, considerably, the class of operations given in section 2.0. Ultimately, one must return to the queueing problems created by these operations.

5.3 Some Relevant Literature. Disney [1975] is the only overall review, with which we are familiar, of results concerning this approach to queueing networks. Of particular interest would be the forthcoming book by Franken, et al. [1980] concerned with properties of marked point processes in a queueing framework. The work of Çinlar [1972] is a splendid review of the superposition problem in a point process setting. The departure process problem is nicely reviewed in Daly [1976]. We know of no review of the problem of random deletions. The paper of Daly and Vere Jones [1972] is a good introduction to stochastic point process theory, random deletions and random stretching.

There are a great many loose ends in these studies of non-Jackson networks. In most cases much new work has been done on each of the topics covered by the cited reviews. Little, if any, of the work is as coherent and tidy as the work on Jackson networks. As a consequence one must search diligently over a broad spectrum for results on these many related topics.

6. Summary

6.0 Summary. We have attempted to review in a few pages, more than 20 years worth of research in the field of queueing network theory. To accomplish this in such a short space we have concentrated on two topics: Jackson networks and flow in networks. Our primary emphasis has been on the Jackson network results. In areas of application these are the results that are of primary importance. Under the pressing restrictions of time and space we have concentrated only on the basic Jackson work and the important Kelly work. While we have alluded to other work, we have by no means provided a definitive state-of-the-art survey. Considerable work has been done on Jackson networks in the past 15 years. This work is to be found principally in the literature of the computer scientist whose interests in these topics seems to have revitalized that field. We can only hope that the reader interested in a host of results and fascinating applications will consult this literature. The best starting point is probably the two volumes of Kleinrock [1975; 1976] and especially the interesting chapters 4, 5, 6 of volume II which present some of the basic queueing problems occurring in computer networks as well as an interesting discussion of trials and tribulations of applying known results to the design of a large scale system. Beyond that we can only suggest that the interested reader peruse the journals of computer science (e.g., J.A.C.M. or Acta Informatica) as these topics continue to be researched and applied.

The study of flow processes in queueing networks is fragmented at present. There are many results. We have mentioned a few. There is much that we have not said and much that remains to be said. It is our view that the link up with the more general field of marked point process theory is natural. For the study of flow processes in networks, it is natural to think of the network and its components as operators on random point processes as discussed in section 5. That view will probably provide greater generality and deeper insights into these flow processes than is now possible.

There are many other topics in these areas that we have not even mentioned. The useful computational work of Wallace [1974] and Neuts (for example, Neuts [1979]) has been left untouched. The study of closed networks has been ignored (e.g., Gordon and Newell [1967]). The interesting network decomposition idea of Courtois [1978] deserves attention both for its theory as well as its application potential. The concepts of approximations, including diffusion approximations and heavy traffic approximation, have not even been mentioned. One might consult Harrison [1978] to start into this area. It would seem that there is no end to such an enumeration. The study of queueing networks is an enormously large and diverse field. Our tutorial has at best "hit the high spots".

Acknowledgement

I would like to thank Robert D. Foley and Burton Simon for many helpful discussions in the preparation of this paper.

References

This list of references is intended to be a guide to the literature. If an author wrote two papers, one a continuation of the other, we only reference the latter under the knowledge that the first paper is included in the bibliography of the second. In this way the references can be used to get one started in the field. More extensive searching would then have to be done by the usual "follow your nose" principle.

1. Barbour, A. D. (1976), "Networks of Queues and the Methods of Stages," Adv. Appl. Prob., 8, 584-591.

2. Borovkov, A. A. (1976), Stochastic Processes in Queueing Theory, Springer-Verlag, New York,

3. Brockmeyer, E., Halstrom, H. L., and Jensen, A. (1948), "The Life and Work of A. K. Erlang," Trans. Danish Acad. Tech. Sci., Transactions No. 2.

4. Burke, P. J. (1956), "The Output of a Queueing System," Oper. Res., 4, 699-714.

5. Burke, P. J. (1964), "The Dependence of Delays in Tandem Queues," Ann. Math. Stat., 35, 874-875.

6. Burke, P. J. (1969), "The Dependence of Sojourn Times in Tandem M/M/s Queues," Oper. Res., 17, 754-755.

7. Çinlar, E. (1972), "Superposition of Point Processes," Stochastic Point Processes: Statistical Analysis, Theory, and Applications (ed. P.A.W. Lewis), Wiley, New York.

8. Courtois, P. J. (1977), Decomposability: Queueing and Computer Systems Applications, Academic Press, New York.

9. Cox, D. R. and Smith, W. L. (1961), Queues, Chapman and Hall, London.

10. Daley, D. J. (1976), "Queueing Output Processes," Adv. Appl. Prob., 8, 395-415.

11. Daley, D. J. and Vere Jones, D. (1972), "A Summary of the Theory of Point Processes," Stochastic Point Processes: Statistical Analysis, Theory, and Applications (ed. P.A.W. Lewis), Wiley, New York.

12. Disney, R. L., Farrell, R. L., and de Morais, P. R. (1973), "A Characterization of M/G/1/N Queues with Renewal Departures," Mgmt. Sci., 20, 1222-1228.

13. Disney, R. L. (1975), "Random Flow in Queueing Networks: A Review and Critique," Trans. Amer. Inst. Industr. Engr., 7, 268-288.

14. Disney, R. L., McNickle, D. C., and Simon, B. (1980), "The M/G/1 Queue with Instantaneous Bernoulli Feedback," (to appear, Nav. Res. Log. Quart., Dec.).

15. Disney, R. L. (1978), "Sojourn Times in Queues with Feedback," paper presented at Colloquium on Point Processes and Queueing Theory, Keszthely, Hungary, Sept. 4-8, 1978. To appear in Proceedings of Keszthely Conference.

16. Feller, W. (1966), An Introduction to Probability Theory and Its Applications, vol. 2, Wiley, New York.

17. Franken, P., Konig, D., Arndt, U., and Schmidt, V. (1980), Queues and Point Processes, Akadamie-Verlag, Berlin (to appear).

18. Gordon, W. J. and Newell, G. F. (1967), "Closed Queueing Systems with Exponential Servers," Oper. Res., 15, 254-265.

19. Gordon, W. J. and Newell, G. F. (1967), "Cyclic Queueing Systems with Restricted Length Queues," Oper. Res., 15, 266-278.

20. Harrison, J. M. (1978), "The Diffusion Approximation for Tandem Queues in Heavy Traffic," Adv. Appl. Prob., 10, 886-905.

21. Jackson, J. R. (1957), "Networks of Waiting Lines," Oper. Res., 5, 518-521.

22. Jackson, J. R. (1963), "Jobshop-like Queueing Systems," Mgmt. Sci., 10, 131-142.

23. Kelly, F. P. (1976), "Networks of Queues," Adv. Appl. Prob., 8, 416-432.

24. Kelly, F. P. (1979), Reversibility and Stochastic Networks, Wiley, New York.

25. Kleinrock, L. (vol. 1, 1975, vol. 2, 1976), Queueing Systems, Wiley Interscience, New York.

26. Melamed, B. (1979), "Characterizations of Poisson Traffic Streams in Jackson Queueing Networks," Adv. Appl. Prob., 11, 422-438.

27. Neuts, M. F. (1978), "Markov Chains with Applications to Queueing Theory, which have Matrix Geometric Invariant Probability Vector," Adv. Appl. Prob., 10, 185-212.

28. Prabhu, N. U. (1965), Queues and Inventories, Wiley, New York.

29. Reich, E. (1957), "Waiting Times when Queues are in Tandem," Ann. Math. Stat., 28, 768-773.

30. Schassberger, R (1978), "The Insensitivity of Stationary Probabilities in Networks of Queues," Adv. Appl. Prob., 10, 906-912.

31. Simon, B. and Foley, R. D. (1979), "Some Results on Sojourn Times in Acyclic Jackson Networks," Mgmt. Sci., 25, 1027-1034,

32. Syski, R. (1960), Introduction to Congestion Theory in Telephone Systems, Oliver and Boyd, Edinburgh.

33. Takacs, L. (1963), "A Single Server Queue with Feedback," Bell Syst. Tech. J., 505-519.

34. Wallace, V. L. (1974), "Algebraic Techniques for Numerical Solution of Queueing Networks," Math. Methods in Queueing Theory, Lecture Notes in Economics and Mathematicsl Systems, No. 98, Springer-Verlag, New York.

35. Witt, W. (1974), "The Continuity of Queues," Adv. Appl. Prob., 6, 175-183.

DEPARTMENT OF INDUSTRIAL ENGINEERING AND OPERATIONS RESEARCH
VIRGINIA POLYTECHNIC INSTITUTE AND STATE UNIVERSITY
BLACKSBURG, VIRGINIA 24061

[illegible] 302–311.

[illegible]
[illegible]

Proceedings of Symposia in Applied Mathematics
Volume 25, 1981

PRACTICAL ASPECTS OF FISHERY MANAGEMENT MODELING

Frederick C. Johnson[1]

ABSTRACT. The Pacific Coast salmon fisheries are considered to be the most sophisticated fishery system in the world [1]. In spite of a great deal of theoretical work [2], [3], [4], the practical problems associated with the management of these fisheries require a level of detail far exceeding a general theoretical framework for optimal resource utilization. While many of the problems are associated with the characteristic nature of common property resources [5], there also exist special catch allocation problems arising from federal court decisions on treaty Indian fishing rights [6]. In this paper we will first give an overview of the technical and other problems arising in salmon fisheries management. We will then discuss the development of a mathematical model for fishing regulation analysis. This model has become a standard tool used by the Washington State Department of fisheries to evaluate the economic and biological impacts of alternative salmon fishery regulation policies.

1. INTRODUCTION

The scientific management of any high seas fishing industry is faced with many complex technical, political and economic problems. These technical problems arise from the basic nature of fishing and fish finding which makes the determination of fundamental biological parameters such as natural mortality rates and stock size a difficult and generally imprecise process. The political and economic problems associated with fishing stem from the fact that fishing is the sole remaining major food production system which relies on the hunting and exploitation of wild stocks that are a common property resource. As a result of the common property nature of a fishing resource, the conservation, control and allocation of the resource is a highly sensitive political problem, and agreement among competing users on a unified management policy is difficult to achieve. The situation is further aggravated by the above mentioned technical problems which often result in a lack of scientific consensus

---

1980 Mathematics Subject Classification. 65C20, 68J05, 92A15.
[1]Supported by the Washington State Department of Fisheries.

on underlying biological parameters, and this lack of consensus provides a ready made reason to maintain the status quo and to deemphasize important long term issues.

The common property nature of the fisheries also leads to economic problems for fishermen due to the phenomenon that good fishing tends to create more fishermen. The end result can be a seriously overcapitalized fishery in which only a fraction of the fishing gear is required to harvest the resource with economic efficiency. Such excess fishing capacity causes severe stress in fishery regulation for two reasons. Without careful management, the fish stocks may be severely overharvested, even to the point of extinction. Also, the practical politics of an overcapitalized situation often makes it more expedient to restrict the efficiency of the harvestors, by season limitations or by limiting the types of fishing gear which may be legally used, than to restrict the number of harvestors. Such limitatations can be highly controversial.

A final crucial feature of the common property nature of fisheries is that almost all resource management must be achieved by external regulation rather than by self-control. The competition for the resource among fishermen means that it is generally not in the best self-interest of a fisherman to practice conservation since most of the benefits of his efforts accrue to his competitors rather than directly to himself.

The Pacific Coast salmon fisheries exhibit all of the problems of a common property resource. In addition, the State of Washington is faced with special management problems of mandated catch allocation among certain user groups. A federal court ruling in 1974 (upheld by the United States Supreme Court in 1979) stated that Indian treaties signed in the 1850's gave the treaty tribes the right to harvest fifty percent of the salmon originating in Puget Sound and other areas of Washington State. Since those tribes had been catching only about five percent of these fish, the development of new regulations to meet the court requirements was a major task. Because of the large number of alternative regulation sets that seemed feasible, the many different salmon stocks involved and the complex interrelations between stocks and fisheries, the development of a mathematical model for salmon fishery management was selected as the most suitable approach to this regulatory task. The resulting model has subsequently become a standard management tool in Washington State.

A major goal of the fishery model was to make maximum use of available data to provide a level of detail adequate to meet management needs. Fortunately, radically improved fish tagging techniques were available to provide more information on stock composition of catches than was previously available. A special challenge was to make good use of these new data.

As will be seen from the subsequent discussion of the model, the basic equations contained in it are straighforward. The success of the model is based not on the derivation of complex equations but on the ability to couple large amounts of data in a systematic analysis which provides useful results for resource managers. This was achieved by working closely with key biologists over a period of many months to carefully define the type and quality of available data, the identify management questions that must be answered by the model and considering many alternative model configurations. Because of this lengthy analysis phase and the deep understanding of the biologists, the subsequent implementation of the model was achieved without major technical problems.

2. THE SALMON FISHERIES

Salmon are anadromous fish. This means that they reproduce in fresh water but spend their adult lives in the ocean. Although salmon are highly migratory and can range over several thousand miles, they have a highly developed homing instinct, and the vast majority return to their natal streams to spawn. Unlike the Atlantic salmon all Pacific salmon die after spawning. Because of the strong homing characteristics of the salmon, the concept of a salmon stock is fundamental to management. A stock of salmon is a subset of a particular species which has the same stream of origin. Ideally, each stock of salmon is individually managed to achieve its escapement goals, that is, to obtain the required number of spawning fish to reproduce the stock.

This is a fundamental difference in management from other ocean species. Because appropriate spawning grounds are limited and fixed, a relatively accurate determination can be made of escapement requirements. Since there is only a certain amount of space for egg deposition, overescapement does not produce more juvenile fish, it results in a waste of the surplus adults. Species which spawn in the ocean, on the other hand, do not have such a restricted spawning potential and important questions arise as to the appropriate number of adults that should be unharvested and left to spawn each year.

Salmon originating in the State of Washington are harvested in sport and commercial fisheries from Alaska to northern California. Because of the highly migratory nature of the salmon, regulation changes in one fishing area impact all of the fisheries throughout the range of the fish. In essence, a regulation change resulting in an increase (or decrease) of catch in one fishery results in a decrease (or increase) of catch in the other fisheries in the migration path of the salmon. Therefore, all of the fisheries must be treated simultaneously when analyzing proposed regulatory changes.

The management of these fisheries is faced with two other important factors: hatchery stocks and mixed stock fisheries. Because of the loss of natural spawning grounds due to logging, construction of dams and urbanization in general, salmon hatcheries are used as a major source of fish for the sport and commercial fisheries. A key feature of hatcheries is that natural mortality, between the taking of eggs and the release of juvenile fish, is substantially reduced. Consequently far fewer adult fish are needed to meet the spawning requirements of a hatchery stock as compared to a wild stock of equal size, and the hatchery stock can be much more heavily harvested. In other words, the harvest rate of a hatchery stock can be much higher than for a wild stock, and this rate disparity is a major concern wherever wild and hatchery stocks are simultaneously harvested. Since mixed stock fishing occurs in many areas, the effects of changes in regulations on escapement needs are vitally important.

3. THE FISHERY MODEL

The fishery model was developed to serve as a management tool for analyzing the economic and biological effects associated with changes in salmon fishery regulations. The goal of the model is to provide a common methodology for quantifying the impact of alternative sets of fishing regulations on fisheries performance and salmon stock abundance and stability. The performance of the salmon fisheries is determined by performing a detailed calculation of the redistribution among the ocean and inside fisheries in Puget Sound and the Columbia River of catch, weight and catch value that is a consequence of each proposed set of fishery regulations. The evaluation of stock abundance and stability is based on an escapement analysis which determines whether specific escapement goals are met on a stock by stock basis. Input data for the model are used to calculate detailed fishing rates, stock recruitment populations and other factors such as maturation rates and length/weight relationships. All of these parameters are then combined with new fishery regulation data to evaluate the effects of regulation changes.

Regulation evaluation is based on a complete, time-sequenced simulation of the interactions between the fisheries and the salmon life cycle. This simulation begins with the initial recruitment of the salmon stocks to the fisheries, continues through the entire ocean life of each stock, and ends with the final escapement of the oldest fish. From initial recruitment to final escapement, the model simulates both the effects of fishing and the biological process of growth, natural mortality, maturation, and migration. The effects of fishing are based on the population size and distribution of the stocks, their length characteristics, the regulations controlling the fisheries, and the historical performance of the fisheries as reflected in catch statistics. The total catch, weight, and catch value is obtained for each fishery, and the total escapement is obtained for each stock. This information forms the basis of comparison between alternative fishery regulations.

The evaluation of fishing regulation changes is ultimately dependent upon how well alternative regulations satisfy a management policy which consists of catch distribution and escapement goals. Because of the frequently complex nature of this evaluation, no attempt was made in the fishery model to compile an overall measure of worth for each set of regulations. The model simply computes the effects of regulations--final evaluation remains in the hands of the management agency.

The fishery model is a steady-state model in which the fishery regulations, effort, and other characteristics are held constant throughout the life of the stock. This type of analysis can be viewed as giving the representative effects of a set of regulations averaged over several years. Such a steady-state analysis is appropriate for evaluating regulations from a policy viewpoint. The stochastic components of the stock/fishery system, such as recruitment failures, bonanza runs, and effort fluctuations, generally cannot be forecast with sufficient accuracy to be considered from a policy viewpoint. These variations are more appropriately treated as special cases which (possibly) require special regulations. In addition, this analysis does not indicate the effects of the transition period in which a brood year of a stock is fished under both old and new regulations. The fishery model measures the effects of a set of regulations after this transition phase has passed.

It is important to understand that the objective of the fishery model is not to predict what will happen in any specific year. Rather, the purpose of the model is to serve as a tool for the evaluation and comparison of alternative fishery management policies. The model provides a common methodology and computational basis for regulation testing, and it is the comparisons between different sets of regulations under a common set of computational assumptions that provides fishery managers with insight into the strengths and weaknesses of proposed regulation changes. In effect, the model serves as a computational laboratory for regulation analysis. The challenge for the biologists and mathematicians during the development of the model was to combine data, equations and knowledge gained through experience into a single tool which could provide useful quantitative information on the effects of regulation change.

The fishery model is concerned with two principal entities, salmon stocks and salmon fisheries. It deals with the stocks of two salmon species, Chinook and Coho. In addition to being a particular species further classified by spawning ground location, a stock may also be thought of as a group of fish having a common tag or mark. The term salmon fishery can refer to a collection of individual fisheries or to a specific fishery. Each fishery in the model is identified by the location of the fishing activity, the type of fishing gear used, the target species of salmon, and the participants (users) in the fishery. This is necessary since different participants can be subject to different sets of regulations.

The fishery model contains two principal phases--calibration and regulation analysis. The calibration phase is the major portion of the model and is a complex process. After a calibration has been performed, however, the resulting parameters are used to analyze alternative management policies by multiple executions of the analysis phase. Recalibration is necessary only when new stock, fishery, or catch data become available.

4. BASIC CATCH EQUATIONS

The fishery model is based on a set of equations relating to fishing and natural mortality due to Beverton and Holt [7]. Let $N(t)$ be the population of a stock at time t. Then $N(t)$ is assumed to satisfy the differential equation

$$\dot{N}(t) = -ZN(t)$$

$$Z = M + F$$

where M is the natural mortality rate and F is the fishing mortality rate. Let t_0 be an initialization time and let $N_0 = N(t_0)$. Then we have

$$N(t) = N_0 e^{-Z(t - t_0)} \tag{1}$$

and

$$C(t) = \frac{F}{Z}(N_0 - N(t)) \tag{2}$$

where $C(t)$ is the total catch in the time period $(t - t_0)$. When more than one fishery is active, we have

$$F = \sum_j F_j$$

and

$$C_j(t) = \frac{F_j}{Z}(N_0 - N(t))$$

where F_j is the fishing rate in fishery j and $C_j(t)$ is the corresponding catch.

If we adopt a uniform time interval $\Delta t = 1$ and let $N_i = N(t_0 + i\Delta t)$, equations (1) and (2) can be written in the sequential form

$$N_{i+1} = N_i e^{-Z_i} \tag{3}$$

$$C_i = \frac{F_i}{Z_i}(N_i - N_{i+1}) \tag{4}$$

where the mortality rates M_i and F_i are allowed to vary with i, and C_i is the catch in the time period $(t_0 + i\Delta t)$.

When the time period Δt is small (Δt is one month in the model), equations (3) and (4) can be linearized to give

$$N_{i+1} = S_i N_i - F_i N_i \tag{5}$$

$$C_i = F_i N_i \tag{6}$$

where $S_i = 1 - M_i$ and is the survival rate. Equations (5) and (6) are the basic catch equations of the model.

Once N_0, $\{F_i\}$ and $\{S_i\}$ are known, equations (5) and (6) can be used to compute the catch of the stock in all fisheries. While $\{S_i\}$ can be obtained from external studies, neither N_0 nor $\{F_i\}$ are known and must be derived from a calibration process. This holds true even for a hatchery stock for which a relatively accurate estimate of the number of fish released is available. The reason is that juvenile salmon enter the ocean at a very small size and essentially 'disappear' for a year until they have attained sufficient growth to be caught on standard fishing gear. No good estimates for natural mortality exist for this period, but since the fish are small and more subject to predation, it is much higher than for adult salmon.

Returning to the calibration process, let $\{C_i\}$ be an historical catch record for the stock. We can include in this data set all escapement data since they are readily available. In fact, without loss of generality, escapement can be treated as just another fishery. Since all salmon die after spawning, the solution to the calibration problem is to rewrite (5) and (6) in a backwards time fashion,

$$N_i = (N_{i+1} + \overline{C}_i)/S_i \tag{7}$$

$$F_i = \overline{C}_i/N_i \tag{8}$$

and to initialize these equations with $N_L = 0$, where T_L is the time period after the oldest adult has spawned and died. Thus, the catch record and estimates for the survival rates are adequate to derive both catch rates and the initial stock population.

5. KNOWNS AND UNKNOWNS

While equations (5)-(8) provide the overall framework for the model, many important details remain to be added. First, however, we need to have a clearer understanding of which information is known and what must be computed.

There are three categories of input data for the calibration phase of the model:

Fishery specifications
Stock specifications
Historical catch data

The fishery specifications describe the physical and economic characteristics of each salmon fishery, and the stock specifications describe the biological characteristics of each salmon stock. The catch data describe the effects of the stock-fishery interactions.

The following data are used to specify each salmon fishery:

1. area,
2. target species,
3. participant type,
4. gear type,
5. season specifications,
6. induced mortality data,
7. catch values.

The area specifies the location of the fishery, and the target species specifies the species of a salmon caught by the fishery. The season specifications consist of an opening date and a closing date for the fishing season and the minimum legal length of the catch during the season. A fishery may have more than one season during the year.

Induced mortality refers to mortality caused by the act of fishing but which does not generate any catch value. Three types of induced mortality are considered: hooking, drop-out and cross-species. Hooking mortality arises in a fishery which has a size-limit since the process of releasing sublegal fish kills a significant fraction. Drop-out mortality applies to net fisheries and occurs when legal sized fish are killed but are not landed because they drop out of the net. Cross-species mortality arises in a mixed species situation in which the season for one species is closed. Fishing for the open season species causes mortality in the closed season species since these fish must be released after landing. Induced mortality data are given as a fraction of the catch, i.e., a 10% rate means that for each 10 legal fish caught one fish is killed due to induced mortality.

The catch values for commercial fisheries are ex-vessel prices in dollars per pound (round weight). Since the dollar per pound price can change based on the total weight of the fish, the price is specified by weight range. In other words, a fish weighing less than 8 lbs. may be worth $0.75 per pound, while a fish larger than 8 lbs. may be worth $1.00 per pound. A similar scheme is used for sport fishery values, but in this case, the value is specified as dollars per fish. Since some difference of opinion exists concerning sport fishery values, this technique can be used to represent any particular sport valuation philosophy.

The stock specification data consist of:

1. species,
2. natural mortality rates,
3. growth data,
4. length frequency data,
5. migration paths,
6. substock ratios,
7. escapement goals,
8. terminal fisheries.

An underlying assumption of the fishery model is that changes in fishery regulations will not have a significant impact on the fundamental growth rate of the salmon stocks. Although changes in size limit will naturally affect the average length and weight of the catch, this is due to a change in the fraction of the population that is legal-sized and not due to growth mechanism changes. Therefore, the fishery model does not use a growth equation, but instead has as input fixed monthly lengths which define the growth curve, and the data are given for each age and maturity combination. It is assumed that the lengths of a population are normally distributed, and the mean of the distribution is given by the monthly length data. The standard deviation is specified by the length frequency data which give the minimum and maximum mean lengths for the stock and the corresponding standard deviations. Linear interpolation is used to obtain intermediate standard deviation values. The length/weight relationship of a stock is obtained by performing a least-squares analysis on catch data to compute the parameters of the standard length-weight equation having the form $W = aL^b$, where W is weight and L is length.

While the stock is the principal salmon entity, a stock must be further categorized into substocks in order to account for major variations in the migration patterns of a stock. All substocks have the same growth and natural mortality characteristics, but differ by migration pattern. The migration data specify the fishing areas over which a substock is distributed, and the migration patterns are distinguished by age and maturity. The substock ratios specify the fraction of the total stock population (at initial recruitment) for each substock. It should be noted that these ratios will change throughout the life of a stock due to the effects of exposure to different fisheries.

The escapement goal for a stock represents the total number of fish required for spawning purposes. Surplus escapement or under-escapement is automatically allocated to or taken from the specified terminal fisheries. Thus, the model treats terminal fisheries as control fisheries which are adjusted to meet escapement goals, and this agrees with management practice. This aspect of terminal net fishery management is a key factor in achieving a steady-state analysis. Unless a stock is so depressed or non-terminal area fishing is so intense that escapement goals are impossible to meet, terminal fishery management insures a constant spawning population. This insures that the initial population size N_0 does not vary from year to year, a fundamental requirement for a steady-state analysis.

The vast majority of the input data to the model consist of catch data. The catch data consist of a number of fish caught, the average length of the fish, and the average weight of the fish. These data are categorized by stock, fishery, age, sex and maturity of the fish, and the time period of the catch. These data are based on current stock size projections as applied to composites of marked fish experiment groups and form the basis for calibrating parameters in the model. Several hundred catch data items may be required for each stock.

Results from recoveries of marked fish experimental groups reflect the interactions of the stocks and fisheries under the specified fishing regulations. Typically, the experimental data are averaged over several years to smooth both survival and effort variations. The calibration phase thus utilizes current stock size projections, composites of marked fish experimental results, and the corresponding fishing regulations to calibrate the model parameters.

It may be noted that escapement data have not appeared explicitly in the above description. As stated earlier, escapement is treated as simply another fishery, and the mortality for this fishery actually represents the number of fish reaching the spawning grounds. Also, there is no explicit specification of maturation rates for a stock. Since the catch data are distinguished by sex and maturity, the maturation rates are derived from these data. Maturation refers to the process of attaining sexual maturity. During each year of life a fraction of the population (different for male and female) attains sexual maturity and begins the homeward migration. The remaining immature fish continue the general outbound migration. Growth, natural mortality and migration patterns are all a function of maturity.

None of the model biological processes of growth, natural mortality, migration of a stock are affected by fishing on the stock or on other stocks, and this is consistent with current knowledge and experience. Consequently, each stock is a completely independent biological unit. The one mechanism which prohibits complete independence for modeling purposes, however, is induced mortality. This is because induced mortality is known only by species and not by stock. This is most easily seen in the cross-species situation, but it applies to hooking and drop-out mortality as well. Consequently, the calibration process must allocate the induced mortality to individual stocks, and this is the most complex task in the model.

6. THE CALIBRATION PHASE

The principal output of the calibration phase of the model consists of initial population sizes, maturation rates and fishing rates for each stock. The calibration process consist of four steps and utilizes a combination of backward and forward computation.

The first step of the process is a back calculation of the catch data to give a first estimate of the initial population for each stock. At the same time initial estimates of the maturation rates are obtained by calculating (by sex and age) the ratio of mature fish of the stock to the total population of the stock. The maturation calculation is based on the January populations for each age of the fish.

Since the catch data are distinguished by stock and not substock, the next step is to allocate catch to substocks in order to account for differences in migration patterns. This is a forward calculation, and the catch is allocated to each substock based upon the proportional representation of each substock in each catch area.

After the first two steps have been completed for all stocks, the allocation of induced mortality to substocks is the next step. Induced mortality is allocated to all appropriate substock-age-maturity-sex combinations, and this allocation is also based on the proportional representation of the substocks in each catch area. As induced mortality is allocated to a substock it too is back calculated, and final estimates of initial population and maturation rates are obtained for each substock.

The final step of the calibration process is the computation of fishing mortality rates for each substock. These rates are obtained by sweeping forward in time and computing the ratio of legal catch to legal sized population and the ratio of induced mortality to either sublegal or legal population depending upon the type of induced mortality. All of the calibration data are then saved for use in the regulation analysis phase.

7. THE REGULATION ANALYSIS PHASE

The catch data used in the calibration process reflect a particular set of fishery regulations in force when the data were obtained. Thus, an analysis of a regulation change first requires a calculation of the effects of the change on the calibrated fishing mortality rates. Generally, a regulatory change will increase or decrease a fishing rate as, for example, when a fishing season is changed. In addition, there may be regulations which affect the amount of fishing effort, or fishing effort patterns may change as a consequence of new regulations. Such effort changes will also affect the calibrated fishing mortality rates. The conversion of effort changes to rate changes is external to the model and is performed by experienced fishery managers. Changes in minimum length regulations, however, do not affect effort directly but alter the split between the legal and sublegal portions of the stock populations. After the regulation changes have been converted to rate changes, the simulation phase of the model performs a complete simulation of the stock/fishery system and calculates the catch, catch value, and escapement obtained based on the new rates.

Input data for the regulation analysis phase of the model consist of new fishing regulations which are to be evaluated. Only changed regulations need be entered. In addition to regulation changes, fishery values and the number of recruits may also be changed.

Three types of regulatory changes can be analyzed: season changes, effort changes and minimum length changes, and these completely cover the available flexibility of salmon fishery management. An effort change is not necessarily accomplished by regulation, but may be an associated change in fishing patterns generated in response to regulatory changes. The model contains no mechanism which attempts to predict changes in fishing patterns and levels of effort. Instead, this is left to the judgment of the fishery manager during data preparation.

The effects of new fishing regulations are obtained by first computing scale factors for the calibrated fishing mortality rates and then performing a complete life cycle simulation. For each substock population the simulation begins with the recruitment of young fish to the fisheries. The age of recruitment can vary, but recruitment generally occurs in the second year of life. Based on the calibrated maturation rates for male and female fish, the substock population is decomposed into two groups: a group of fish that will mature during the current year of life, and a group of fish which will not mature during the current year. The mature fish are then subjected to the monthly processes of growth and natural mortality. At the same time, the fish are subjected to fishing mortality. The fishing mortalities that affect the population depend upon the migration path of the population (this determines which fisheries are actively exploiting the population), the length distribubution of the population, the types of fishing gear, the regulations controlling the fisheries, and the calibrated fishing rates.

Fishing mortality is either induced mortality and/or catch mortality. The value of the catch is determined by the number of legal fish caught, the average weight per fish, and the price structure of the fishery which caught the fish. Since escapement is treated as a type of fishery, escapement mortality actually determines the number of fish which arrive on the spawning grounds.

The same procedure is applied to the immature fish. These fish are subject to (generally) different growth, natural mortality and migration and, consequently, the effects of the fisheries may be substantially different than was the case with the mature fish. At the end of the calendar year, the remaining immature fish undergo another maturation process, and the entire cycle is repeated for the fish which are now 1 year older.

The annual cycle of maturation, migration, growth, and mortality is continued until all fish have matured and final escapement has occurred. This completes the processing of a substock. After all of the substocks of a stock have been processed, an escapement analysis is performed. Any surplus escapement is allocated to the terminal fisheries or, if there is under-escapement, the terminal fishery catch is reduced and fish are allocated to escapement. This allocation is done on a proportional bases to maintain the correct age ratios.

After all stocks have been processed, the total catch, weight, value, and induced mortality are known for each stock/fishery combination, and the total escapement for each stock has been obtained. These data are summarized and reported by the model, and this completes the regulation analysis phase. Following a single brood year throughout its life cycle in the fisheries is, under steady-state conditions, equivalent to harvesting multiple age classes in a single year. Thus, the output of the model gives a complete picture of a single year of fishing.

8. FINAL REMARKS

Questions concerning validation and accuracy have no simple answers for the fishery model. Appropriate historical data are not available to use as a basis of comparison with current data for classical validation purposes. Consequently, test and validation of the model utilized a series of simplified test cases and parametric studies. The behavior of the model using full-scale data has been carefully examined by management biologists and judged to be realistic and adequate for their needs. It is simply not possible to go beyond this level of validation at the present time. A long term goal for validation is to evaluate the model's behavior using two different sets of steady-state catch data. The current historic record provides one set, but fishery regulations must stablize before a second set of data can be obtained. It is highly likely, however, that this will not occur for several years.

The fishery model has been used by the Washington State Department of Fisheries to analyze proposed salmon fishery regulations since 1976. Because of the very large impact of the court mandated catch allocation and also because of important stock conservation issues, fishing regulations have been changing very rapidly and have received close scrutiny in both public hearings and court proceedings. As a part of this process, the model has been used to analyze as many as 50 different sets of new regulations and variants each year. The usefulness of the model results from incorporating a level of detail that fully utilizes all available data for the examination of complex and conflicting management issues. This was possible because of the excellent understanding of available data, biological mechanisms and management needs that was provided by the management biologists of the Department of Fisheries.

The current version of the model has the capacity to simultaneously analyze 60 stocks, 145 substocks, 190 fisheries in 50 catch areas and 5 different age classes of catch. This represents approximately a doubling of the capacity of the original model. The model undergoes a more or less continuous evolution as additional features, particularly output summaries and graphics are added. There has been no change, however, in the overall design or philosophy of the model, and all versions have been upwards compatible. The current version is expected to be adequate for management needs over the next several years. A more complete discussion of the mathematics of the model may be found in [8]. An example of the complexity of salmon management issues and the role of the model is given in the environmental impact statement and fishery management plan for the 1978 fishing season [9].

BIBLIOGRAPHY

1. P. A. Larkin, "Maybe you can't get there from here: A foreshortened history of research in relation to management of Pacific salmon," J. Fisheries Res. Board of Canada, 36 (1979), 98-105.

2. C. Clark, Mathematical Bioeconomics: The Optimal Management of Renewable Resources, Wiley, New York, 1976.

3. __________, "Mathematical Models in the Economics of Renewable Resources," SIAM Rev. 21 (1979), 81-99.

4. J. Gulland, The Management of Marine Fisheries, Univ. Washington Press, Seattle, Washington, 1974.

5. G. Hardin, "The tragedy of the commons," Science, 162 (1968), 1243-1248.

6. R. Barsh, The Washington Fishing Rights Controversy: An Economic Critique, Monograph Series, Univ. Washington Graduate School of Business Administration, Seattle, Washington, 1977.

7. R. Beverton and S. Holt, "On the dynamic of exploited fish populations," Her Majesty's Stationery Office, London, 1957.

8. F. C. Johnson, "A model for salmon fishery regulatory analysis," NBSIR 75-745, National Bureau of Standards, Washington, July 1975.

9. Anon., "Final environmental impact statement and fishery management plan for commercial and recreational salmon fisheries off the coasts of Washington, Oregon and California commencing in 1978," U. S. Department of Commerce, National Marine Fisheries Service, March 1978.

CENTER FOR APPLIED MATHEMATICS
NATIONAL BUREAU OF STANDARDS
WASHINGTON, D. C. 20234

Proceedings of Symposia in Applied Mathematics
Volume 25, 1981

SOME MATHEMATICAL MODELS IN HEALTH PLANNING

William P. Pierskalla[1]
University of Pennsylvania

ABSTRACT. There are many areas in which mathematical models are useful in health care delivery. Two of these will be discussed here: 1) Diagnostic screening for early detection of disease, and 2) Planning regional blood banking systems.

Non contagious diseases arise in a population in a seemingly stochastic manner. If testing procedures exist which are capable of detecting the disease before it would otherwise become known, and if such early detection provides benefit, the periodic administration of such a test procedure to the members of the population, that is, a mass screening program, may be advisable. Moreover, if the population is composed of sub-populations which exhibit different disease incidence rates and different unit costs of test applications, and if different tests which have different reliabilities for detecting the disease are available, then the question of allocating limited screening resources among the sub-populations arises. The optimal allocation depends upon the form of the disutility functions of the sub-populations. Comprehensive analytic models are needed to perform this allocation.

Health planning can be viewed from many perspectives. Perhaps the most critical one facing the United States today is to contain the costs of health care and yet deliver quality care to the entire population of the U.S. Certain aspects of planning to achieve these objectives must be undertaken on a regional level, others at a sub-regional level, and still others at the institutional level. An integrated hierarchy of analytical models is needed to link the decisions at each of these levels. Decisions at the macro level involve the appropriate numbers of people by skills, numbers of facilities, and technological sophistication for a region. At the middle level, the decisions involve facility locations, their levels of technology and services and personnel needs to achieve minimum cost yet provide accessibility and quality of care in the sub-regions. At the institutional or micro level, analytic models are used to determine admissions and appointments, inventory levels and capital equipment, daily and weekly staffing, and facility scheduling. These different levels of modeling are illustrated in the context of planning for regional blood-bank systems.

---

[1] 1980 Mathematics Subject Classification 90-02, 90B05, 90B25.
This work was partially supported by the National Science Foundation under grant ENG77-07463 and by Office of Naval Research Contract N00014-75-C-0797 (Task NR042-322).

I. DIAGNOSTIC SCREENING FOR EARLY DETECTION OF DISEASE

1. INTRODUCTION

There are many situations where a defect can occur randomly among the members of a population -- either a population of human beings, inanimate objects, or perhaps livestock -- and once present, exist and develop, at least for a time, without any manifest symptoms. If the early detection of such a defect provides benefit, it may be worthwhile to employ a test capable of revealing the defect's existence in its earlier stages. (A defect, disorder, or disease will generally be referred to simply as a defect; and the word unit or individual will refer to a member of the population.)

Of course, continuous monitoring would provide the most immediate such revelation. But considerations of expense and practicality will frequently rule out continuous monitoring so that a schedule of periodic testing -- a screening program -- may be the most practical means of achieving early detection of the defect. In general terms, the question then becomes one of how best to trade off the expense of testing against the benefits to be achieved from detecting the defect in an earlier stage of development. The expense of testing increases both with the frequency of test applications and with the cost of the type of test used. The benefits increase with the frequency and the quality of test used and the quality of a test is often directly related to its cost.

The benefits of early detection also often depend upon the application considered. For example, in a human population being screened for some chronic disease (cancer, glaucoma, heart disease, etc.) the benefits of early detection might include an improved probability of ultimate cure, diminished time period of disability, discomfort, and loss of earnings, and reduced treatment cost. If the population being screened consists of machines engaged in some kind of production, the benefits of early detection might include a less costly ultimate repair and a reduction in the time period during which a faulty product is being unknowingly produced. If the population being screened consists of machines held in readiness to meet some emergency situation, an early detection of a defect would reduce the the time the machine was not serving its protective function. This process of inspecting a sizable population for defects is called mass screening.

The expense of testing includes easily quantifiable economic costs such as those of the labor and materials needed to administer the testing. However, there can also be other important cost components which are more difficult to quantify. For example, in the case of a human population subject to medical screening, the cost of testing includes the inconvenience and possible discomfort necessitated by the test; the cost of false positives which entails both emotional distress and the need to do unnecessary follow-up testing; and even the risk of physical harm to the testee, e.g., from the cumulative effect of X-ray exposure.

The rationale for constructing a mathematical model of mass screening is to provide a conceptual framework within which a mass screening program might best be designed and its worth evaluated. Such a design must determine which kind of testing technology will be used. Several candidate technologies may be available, each with different reliability characteristics and costs. In addition, the frequency of testing must be decided. If the target population can be partitioned into sub-populations according to susceptibility to the defect (e.g., by age, family background, time since last overhaul (for a machine), etc.), then the best allocation of the testing budget among the sub-populations must be determined. Lastly, a decision must be made, for the population and type of defect under consideration, whether a mass screening program is justified at all. It is felt that the above determinations are best carried out within the conceptual framework of a cogent model of mass screening.

In the following sections a reasonably comprehensive model for decision making with regard to mass screening is presented. The model utilizes a time-based approach. But rather than seeking to analyze or minimize detection delay, as in some of the literature on this topic, the objective function is an arbitrary increasing function of detection delay. (Detection delay is the time between the incidence of the defect and its detection, regardless of whether that detection is the result of a screening test or of the defect becoming self-evident.) The reason for choosing a general function is that the disutility experienced upon the delayed detection of a defect may well vary in a highly nonlinear way with the length of the delay.

Another objective sometimes used for inspection models is to maximize the lead time where the lead time is defined as the difference between the time of detection via a screening test and the time detection would otherwise have occurred had not a screening program been in existence. However, in the case in which the test being used is perfectly reliable then it becomes possible to consider the lead-time criterion in terms of the following simple function of detection delay. Suppose that at the (possibly random) age T (measured from the time of incidence of the defect) the defect, in the normal course of its development, would become manifest even without a screening test. Then, if the defect is detected at age t, the lead-time gained is $(T - t)^+$.[1] Therefore, if sup T is finite, the function $D(t) = \sup T - E(T-t)^+$ may be used as a disutility function which, when minimized, will maximize the expected lead-time.

Consequently, the results discussed here based on an arbitrary disutility function, generalize as well as extend results in earlier work. In addition, new results are presented.

In the next section a brief review of the literature is given. Since the work on inspection models is very large, only the most relevant papers are discussed. The interested reader may refer to the surveys by McCall [1965] and Pierskalla and Voelker [1976] for a more comprehensive review.

In the third section the model is formally stated and an expression derived for the expected disutility per unit time incurred under a regime of uniformly spaced test applications.

This model can be specialized to the case of a perfect test; i.e., when the test is administered to an individual with the defect, the defect will be detected with certainty. Under this assumption, a regime of uniform test intervals is optimal within a wider class of "cyclic" testing schedules for a single sub-population. The problem of allocating a screening budget among various sub-populations can then be analyzed. As part of this analysis, the explicit screening schedule $(r_1, r_2, \ldots, r_Q)$, where r_j designates the optimal testing frequency for members of the jth sub-population, can be obtained in terms of the sub-population-specific incidence rates $N_j\lambda_j$ and the budget constraint for the disutility function $D(t) = at^m$. For $D(\cdot)$ convex, the above solution (when m = 1) provides a bound on the ratios r_i/r_j.

Other uses of the model involve the analysis of the case when the probability of detection is a constant over all values of elapsed time since incidence. The long-run expected disutility per unit time can be derived; its differential qualities (with respect to variations in the test reliability parameter and testing frequency) exhibited; and explicit solutions provided for various special forms of $D(\cdot)$. Also for special forms of $D(\cdot)$, decision rules can be presented to select between two different kinds

---

[1] The function $u^+ = \max(u, o)$.

of tests which differ with respect to their reliabilities and cost per application.

If it is assumed that the test will detect the defect if and only if the elapsed time since incidence exceeds (or equals) a critical threshold $\hat{T}$ which characterizes the test, an expression for the long-run expected disutility per unit time can be derived. Then decision rules can be developed to select between two alternative test types which differ with respect to their critical threshold $\hat{T}$ and their cost per application. Some interesting examples for linear and quadratic disutility are given in the fourth section.

The last section in Part I provides a technique to estimate the shape of the disutility function from empirical data in the case of a perfect test.

Finally, it should be mentioned that the population (or each sub-population, in the instances where a heterogeneous population is considered) is assumed to be of fixed size, N, and the defect to arise according to a stationary Poisson process with rate $N\lambda$ (N could be a very large but finite number). It may be somewhat more realistic to set the defect occurrence rate (sometimes called the defect arrival rate) proportional to the number of defect-free units, rather than to the total number of units in the population. However, it is also assumed that no defect can remain undetected longer than T^*, even without any screening tests being given. Hence, $T^*N\lambda$ is an upper bound on the expected number of undetected defects in the population. If it is the case that once a defect is detected, the afflicted unit is replaced in the population with a healthy unit, then $T^*N\lambda$ represents a bound on the expected difference between the number of healthy units and the total number of units in the population. For λ small, as would be the case for a relatively infrequently occurring disorder, this should represent no difficulty. Consequently, it is assumed that λ is small relative to N, which is consistent with the examples of potential applications which have been or will be mentioned.

2. LITERATURE REVIEW

Some of the early papers which have a bearing on time dependent models of mass screening are: Derman [1961], Roeloffs [1963, 1967], Barlow, Hunter, and Proschan [1963] and Keller [1974]. Kirch and Klein [1974], whose paper is addressed explicitly to a mass screening application, seek an inspection schedule which will minimize expected direction delay subject to a constraint on the expected number of examinations an individual would incur over a lifetime; the test is assumed perfectly reliable. The point of view adopted here in Part I is similar to that of Kirch and Klein. However, one respect in which the approach here differs from that of Kirch and Klein is that several sub-populations are considered each with its own characteristic incidence rate for the disorder. The idea is then to allocate optimally a fixed screening budget among the sub-populations. Kirch and Klein instead take a longitudinal view. An individual through his lifetime is subject to different probabilities of incurring the defect and a screening schedule is optimized subject to a constraint on the expected number of examinations over a lifetime.

McCall [1969] considered the problem of scheduling dental examinations under the assumption that the time between the incidence of a cavity and the scheduled dental examination controls whether the cavity results in a filling or an extraction. Cavities are assumed to occur according to a Poisson process. As a generalization, he permits the time required for a cavity to become beyond repair (by a filling) to be a random variable.

Lincoln and Weiss [1964] studied the statistical characteristics of detection delay under the assumption that the times of examinations form a renewal process and that the probability of detecting the defect, $p(t)$, is a function of the defect's age, t. They derive equations, similar to renewal type equations, which relate the density functions for the following entities: the probability of detection at a test application ($p(t)$), the time until the defect becomes potentially detectable and from this time the forward recurrence time to the first test, the probability of the event that at a particular time a test occurs and all prior tests had failed to detect the defect, and the detection delay. For the two special cases where $p(t)$ is a constant and where $p(t)$ is exponential, the moments for the detection delay are derived in closed form. For uniform testing intervals (and general $p(\cdot)$), the distribution and moments of the detection delay are computed. For $p(t) = 1$, they solve for that testing schedule which maximizes the time between tests subject to a constraint on the performance of the screening program relative to detection delay. Two such constraints are considered. The first bounds the probability of detection delay exceeding some threshold T. The second bounds the mean detection delay.

THEOREM I.3.1 in the next section, which gives the expected disutility in terms of the test interval and test type, is similar to the objective function given by Lincoln and Weiss although their approach, as outlined above, is different.

Several recent papers have studied the statistical characteristics of the lead-time provided by a screening program -- either a one-shot screen of the population or a periodic screening program. Some of these papers are: Hutchison and Shapiro [1968], Zelen [1971], and Prorok [1973].

In some recent work, Eddy [1978a, 1978b] describes a semi-Markov Chain analysis for the screening and detection of cancer. This modeling effort was undertaken to devise a practical planning tool based in the reality of a breast cancer screening program.

Finally a few authors, Schwartz and Galliher [1975], Thompson and Disney [1976] and Voelker [1976] let both the reliability of the test and the disutility (or utility) of detection be a function of the defect's state rather than of time since the defect's incidence. Although such models are more general and do utilize a general concept of disutility, they have not been amenable to closed form evaluation of expected disutility.

To incorporate random defect occurrences into their models, previous researchers focus upon an individual who will incur the defect. They use the density function for the age when that individual incurs the defect as a fundamental element of their model. Since the density function reflects age-specific incidence rates, a "life time" testing schedule can, thereby, be developed to tailor testing frequency at each age to the probability that the defect will occur at the age.

Our way of modeling the randomness of these occurrences reflects a somewhat different perspective on the mass screening problem. We look through the eyes of a decision-maker charged with intelligently allocating a fixed budget. The time frame over which the allocation must be made is often short compared to a typical life time of a member of the client population. Therefore, the decision-maker does not plan lifetime screening schedules for particular individuals. Instead, he tries to maximize the benefit that can be derived from his available budget over a much shorter planning horizon.

With the problem viewed in this perspective, the random nature of defect occurrences is most naturally modeled as a Poisson process with its parameter determined by the incidence rate of the defect and the size of the population. This approach has proved particularly useful in the following context: if different segments of the client population exhibit different incidence rates, sub-populations can be defined with defect incidence within each of them modeled as a Poisson process with its respective parameter. Then the budget can be so allocated among the sub-populations as to permit appropriate relative testing frequencies (cf. Voelker and Pierskalla [1978]). In this way, age-specific incidence rates can be incorporated into the notion of Poisson defect occurrences. Moreover, factors other than age which affect defect incidence rates (family history, smoking habits, work environment, etc.) can also be incorporated into the model.

3. A MODEL FOR MASS SCREENING

Although most of the definitions are stated prior to their use in each section, listed below are some notation and conventions used as various times throughout the paper.

(1) $$1_{\frac{m}{r}}(t) = 1_{[\frac{m}{r}, \frac{m+1}{r}]}(t) = \begin{cases} 1 & \text{if } \frac{m}{r} \leq t \leq \frac{m+1}{r} \\ 0 & \text{otherwise} \end{cases}$$

(2) $$\prod_{i=1}^{n} x_i = 1 \qquad \text{for } n < 1$$

(3) $$\sum_{i=1}^{n} x_i = 0 \qquad \text{for } n < 1$$

(4) $[x]$ is the largest integer not exceeding x.

(5) The phrase "increasing function" shall mean a strictly increasing function.

As mentioned previously, the objective function will not be to maximize expected utility, but rather to minimize expected disutility. The disutility associated with a particular detection will depend only on the elapsed time since the defect's incidence. The notation $D(t)$ will express the disutility incurred if detection occurs t units of time after incidence. $D(\cdot)$ is assumed throughout this paper to be a nonnegative increasing function.

In those cases where $\sup_{s \geq 0} D(s) < \infty$, there is a natural relation between the notion of a utility function and $D(\cdot)$; namely,

$$U(t) = \sup_{s \geq 0} D(s) - D(t) .$$

$U(t)$ is a decreasing function expressing the disutility avoided by detecting a defect t units of time after its incidence.

Initially, only a single susceptibility class of size N will be considered. The results will then be generalized to an arbitrary number of subclasses. The times of incidence for the defect in the population are assumed to form a Poisson process with parameter $N\lambda$ and are designated by the sequence $\{s^k\}$, $k = 1, 2, \ldots$ This assumption is motivated by the fact that any occurrence or arrival process with the following characteristics is a Poisson process: at the time of an arrival there is almost surely (i.e., with probability one) only one arrival, that the number of arrivals in a time interval does not depend on past arrivals, and that the number of arrivals in intervals of equal length are identically distributed (Cinlar [1975]). For many applications such as the occurrence of diseases like cancer, heart trouble, etc., this assumption is quite reasonable.

Were a screening test of type ℓ to be administered to the individual with the kth defect (i.e., kth in the order of defect incidence) at time $S^k + t$, then the test outcome random variable is:

DEFINITION:

$$Y_\ell^k(t) = \begin{cases} 0 & \text{if kth defect is not detected at time t after incidence} \\ 1 & \text{if kth defect is detected at time t after incidence.} \end{cases} \tag{3.1}$$

Since a defect cannot be detected before its incidence, $Y_\ell^k(t) = 0$ for $t < 0$. Notice that the argument of $Y_\ell^k(\cdot)$ refers to time relative to the incidence of the defect. If the population is screened at time s, where s is absolute time, then the kth defect will be detected if and only if $S^k \leq s$ and $Y_\ell^k(s - S^k) = 1$.

It is assumed that $Y_\ell^k(t)$, $Y_\ell^j(s)$ are independent except when $k = j$ and $t = s$, and that $P(Y_\ell^k(t) = 1)$ depends only on ℓ and t. This latter assumption makes possible the

DEFINITION:

$$p_\ell(t) = P(Y_\ell^1(t) = 1). \tag{3.2}$$

The function $p_\ell(\cdot)$ describes the reliability characteristics of the type ℓ screening test, i.e., how likely such a test is to detect a defect as a function of the defect's age. Note that $p_\ell(t) = 0$ for $t < 0$.

In this section the screening test is assumed to be administered to the entire population at the times 1/r, 2/r, 3/r,... The testing frequency r is a control variable. In the next section, upon assuming perfect test reliability, it is shown that the above schedule of uniform testing intervals is optimal within a wider class of "cyclic" schedules.

The random variable $\bar{S}_{r,\ell}^k$ denotes the time at which the kth defect is detected. $\bar{S}_{r,\ell}^k$ depends on the arrival time of the defect (S^k), the type of test used (ℓ), and the testing frequency (r).

DEFINITION:

$$\bar{S}_{r,\ell}^k = \min_{n=1,2,...} \{n/r \mid Y_\ell^k(\tfrac{n}{r} - S^k) = 1\} \tag{3.3}$$

Given the application of test type ℓ at the times $\{1/r, 2/r, \ldots\}$, the disutility incurred by the kth defect is $D(\bar{S}^k_{r,\ell} - S^k)$. The total disutility incurred due to those defects which occurred in the interval $[j/r, (j+1)/r)]$, when test ℓ is used is:

DEFINITION:

$$B_{r,\ell,j} = \sum_{k=1}^{\infty} D(\bar{S}^k_{r,\ell} - S^k)1_{j/r}(S^k).$$

The following proposition provides an expression for $E[B_{r,\ell,j}]$ in terms only of the disutility due to detection delay (as expressed by $D(\cdot)$) and of the reliability of test type ℓ (as expressed by $p_\ell(\cdot)$). In addition, the theorem shows that $E[B_{r,\ell,j}] = E[B_{r,\ell,0}]$ for $j = 0, 1, 2, \ldots$

THEOREM I.3.1:

$$E[B_{r,\ell,j}] = N\lambda \sum_{n=1}^{\infty} \int_{\frac{n-1}{r}}^{\frac{n}{r}} D(u)p_\ell(u) \prod_{m=1}^{n-1} [1 - p_\ell(u - \frac{m}{r})]du$$

$$\text{for } j = 0, 1, 2, \ldots$$

This proposition yields a relatively simple expression for the expected total disutility in any interval of length $1/r$ and using test type ℓ with probability of detection $p_\ell(\cdot)$. This expectation is used in the objective function of a mathematical program to determine the optimal testing frequency for a mass screening program for a heterogeneous population. In order to develop this mathematical program, the following definitions are useful.

DEFINITION:

$$\bar{B}_{r,\ell} = \lim_{n\to\infty} \frac{r}{n} \sum_{j=0}^{n-1} E[B_{r,\ell,j}].$$

$\bar{B}_{r,\ell}$ is the long-run expected disutility per unit time given the testing frequency r and the test type ℓ. (The factor r enters the definition to convert disutility per unit testing-interval into per unit time.) By *Theorem I.3.1*, $E[B_{r,\ell,j}] = E[B_{r,\ell,0}]$ for $j = 0, 1, 2, \ldots$ Therefore,

$$\bar{B}_{r,\ell} = rE[B_{r,\ell,0}]$$

$$= rN\lambda \sum_{n=1}^{\infty} \int_{\frac{n-1}{r}}^{\frac{n}{r}} D(u)p_{\ell}(u) \prod_{m=1}^{n-1} [1 - p_{\ell}(u - \frac{m}{r})]du. \tag{3.4}$$

Notice that if test type ℓ provides perfect reliability, i.e., $p_{\ell}(t) = 1$ for $t \geq 0$, then (3.4) gives

$$\bar{B}_{r,\ell} = rN\,\lambda \int_0^{\frac{1}{r}} D(u)du. \tag{3.5}$$

Now consider the problem of selecting testing frequencies and test-types for each of Q different susceptibility classes which together comprise the whole population. These classes may differ from one another in the number of units they contain, in their defect incidence intensity, and in the cost per test application to an individual for a particular type of test. The sub-populations, however, are assumed to share a common $D(\cdot)$ function.

Using (3.4) the expected long-run disutility per unit time for sub-population j with frequency $r(j)$ and test $\ell(j)$ where $j = 1, 2, \ldots, Q$ is given by:

$$\bar{B}_{j,r(j),\ell(j)} = r(j)N_j\lambda(j) \sum_{i=1}^{\infty} \int_{\frac{i-1}{r(j)}}^{\frac{i}{r(j)}} D(u)p_{\ell(j)}(u) \prod_{m=1}^{i-1} [1-p_{\ell(j)}(u - \frac{m}{r(j)})]du.$$

This expression can be used to formulate a multi-sub-population screening problem subject to a budget constraint as follows:

$$\underset{(r(1),\ldots,r(Q),\ \ell(1),\ldots,\ell(Q))}{\text{Minimize}} \sum_{j=1}^{Q} \bar{B}_{j,r(j),\ell(j)} \tag{3.6}$$

such that

$$\sum_{j=1}^{Q} N_j c_{j,\ell(j)}\, r_j \leq b \tag{3.7}$$

$$r_j > 0 \qquad j=1,\ldots,Q \tag{3.8}$$

$$\ell(j) \in \mathcal{L} \quad j=1,\ldots,Q \tag{3.9}$$

where

$c_{j,\ell(j)}$ = cost per application of a test of type $\ell(j)$ to an individual of sub-population j,

N_j = number of units (or individuals) in sub-population j,

b = budget per unit time.

$\mathcal{L}$ = set of all feasible tests .

In order to make this mathematical program even more comprehensive, it is possible to add constraints on the amount of testing labor available and on the capacity of the testing facilities in terms of the number of arrivals, the frequency of testing and the type of tests used. In addition, the cost of false positives can be included as a part of the test costs in inequality (3.7).

For example, a constraint on the total labor available is:

$$\sum_{j=1}^{Q} N_j \delta_{j,\ell(j)} r_j \leq L_\ell \tag{3.10}$$

and on the total testing facilities available is:

$$\sum_{j=1}^{Q} N_j f_{j,\ell(j)} \, r_j \leq F_\ell \tag{3.11}$$

for each type of test $\ell(\cdot)$ used over the sub-populations being tested, where

$\delta_{j,\ell(j)}$ = amount of labor needed to administer test type $\ell(j)$ to an individual of population j,

$f_{j,\ell(j)}$ = amount of testing facility time needed to administer test type $\ell(j)$ to an individual of sub-population j,

L_ℓ = total amount of labor available to administer test type $\ell(\cdot)$ per unit time,

F_ℓ = total amount of facility time available to administer test type $\ell(\cdot)$ per unit time.

4. SOME EXAMPLES OF DISUTILITY FUNCTIONS

Suppose a production process is subject to a randomly occurring defect. Although production appears to proceed normally after the incidence of the defect, the product produced is, thereafter, defective to an extent which remains constant until the production process is returned to its proper mode of operation. The only way to learn if the production process is in this degraded state is to perform a costly test. Now, if a test detects the existence of the degraded mode of production t units of time after its incidence, the harm done will be proportional to the amount of defection product (unknowingly) produced which, in turn, is proportional to t. Hence, $D(t) = at$ for some $a > 0$.

Another example where a linear $D(\cdot)$ function may be appropriate would be for the periodic inspection of an inactive device (such as a missile) stored for possible use in an emergency. If t is the time between the incidence of the disorder and its detection, the disutility incurred is proportional to the probability that the device would be needed in that time interval. If such "emergencies" arise according to a Poisson process with rate μ, then the probability of an emergency in a time interval of length t is $1 - e^{-\mu t}$, which, for μ small, is approximately μt. Hence, if b is the cost incurred should there be an emergency while the device is defective, and if μ is the (small) arrival rate of emergencies, then $D(t) = b\mu t$.

A quadratic disutility could arise in the following situation. Suppose the magnitude of a randomly occurring defect increases linearly with time since the occurrence of the defect. For example, the magnitude of the defect might be the size of a small leak in a storage container for a fluid, and as fluid escapes, the leak gets larger. Further, suppose that the harm done accumulates at a rate proportional to the magnitude of the defect. Hence, the quantity of fluid lost (at least initially) increases the longer the defect exists, and the rate of fluid loss is proportional to the size of the leak.

Let the size of the leak (as measured by rate of fluid loss), at a time s since the leak's incidence, be cs. Then, if the defect is detected at time t since incidence, the disutility incurred (fluid lost) is $D(t) = \int_0^t cs\,ds = 1/2\ ct^2$

5. EMPIRICAL ESTIMATION OF D(·)

Although data may not be available currently to support the utilization of particular models, that should not deter the development of such models. It is reasonable to expect that in many application areas in the future much additional data will become available; for example, more data will become available concerning the stochastic pattern of a particular disease's development. Furthermore, a good model will serve as a guide to the kinds of data which should be gathered.

It is also reasonable to expect the future development of improved testing technologies capable of detecting a disorder in a much earlier stage of development than is now the case. Such innovations will make screening programs more attractive and, consequently, make more important the analytical tools to design such programs intelligently.

However, it is possible to provide a simple procedure to estimate empirically the disutility function $D(\cdot)$ under the assumption of perfect detection. Such a procedure is necessary, since upon the detection of a disorder there may be no way of directly determining how long the disorder has been present, although that length of time could not exceed $x = \frac{1}{r}$. It is assumed, however, that at each detection of a defect, the degree of disutility incurred due to that defect can be observed. For example, in the case of medical screening, at the detection of a tumor its degree of development can be noted even if it is not possible to determine exactly when, since the last test, the tumor originated.

In order to allocate optimally a screening budget among differing subpopulations (susceptibility classes), as was discussed in Section 3, it is necessary to know the shape of the disutility function. Under the reasonable assumption that $D(\cdot)$ is an increasing continuous function, the function may be derived in the following manner. For a particular population subject to a Poisson defect arrival $\{S^k\}$ with rate $N\lambda$, arbitrarily select a value for x and set up a prototype mass screening program for the population in which tests are made (using a perfect test) at times 0, x, 2x,... The manner of defects detected at the time jx, which are characterized by a disutility less than or equal to y, record and designate by $Q_j(y)$. Then,

$$Q_y(y) = \sum_{k=0}^{\infty} 1_{[0,y]}(D(jx - S^k))\, 1_{[jx-x,\, jx]}(S^k), \qquad j=1,2,3,\ldots$$

$Q_j(y)$ is observable for all values of disutility $y > 0$. Further, $Q_i(y)$ and $Q_j(y)$ are independent and indentically distributed random variables for $i \neq j$ and $y > 0$. Therefore, by the law of large numbers,

$$\lim_{n\to\infty} \frac{1}{n} \sum_{j=1}^{n} Q_j(y) = E[Q_1(y)] \qquad \text{for all } y > 0.$$

Hence, the function $y \to E[Q_1(y)]$ may be estimated by collecting a sufficiently large sample of the functions Q_1, Q_2,...

Let $H(y) = E[Q_1(y)]$. Once $H(\cdot)$ is estimated in this manner, the following theorem shows how the desired function $D(\cdot)$ may be obtained from $H(\cdot)$.

THEOREM I.3.2: If $D(0) = 0$ and $D(\cdot)$ is continuous and increasing on $[0, x]$. then

$$D(t) = H^{-1}(N\lambda t) \qquad \text{for } 0 \leq t \leq x .$$

By observing the $Q_j(y)$ and taking the inverse of their expectation, an estimate for $D(t)$ may be obtained. For some diseases such as heart disease, glaucoma, and some cancers, there may be enough data currently available to begin estimating the disutility functions and to start computing optimal testing frequencies and tests.

II. PLANNING REGIONAL BLOOD BANKING SYSTEMS

1. A HIERARCHY OF PLANNING MODELS

Although health planning involves many activities and mathematical modeling aspects, we will only discuss hierarchies of models to link decisions at the regional, sub-regional and institutional levels. Decisions at the macro level involve the appropriate numbers of people by skills, numbers of facilities, and technological sophistication for a region. At the middle level, the decisions involve facility locations, their levels of technology and services and personnel needs to achieve minimum cost yet provide accessibility and quality of care in the sub-regions. At the institutional or micro level, analytic models are used to determine admissions and appointments, inventory levels and capital equipment, daily and weekly staffing, and facility scheduling. At all levels it is necessary to make these decisions by trading off competing objectives such as minimizing costs, increasing quality of care, and increasing accessibility and availability of services.

Rather than discuss these modeling activities in an abstract form, which could easily be done, we will focus on regionalization in blood banking.

Blood banks are an important and integral part of health service systems. Their main functions are blood procurement, processing, cross-matching,[2] storage, distribution, recycling, pricing, quality control and outdating. The large blood banks are often also responsible for blood research, disease and

---

[2] Cross-matching is the procedure of testing the donor's blood with a sample of blood from a potential recipient (patient) to determine whether the two types of blood are compatible and therefore will not lead to medical complications when the patient is transfused.

reaction prevention. In recent years, there has been much discussion on the issue of regionalization of blood banking systems, in the hope of decreasing shortages, outdates and operating costs, without sacrificing blood quality, research and education.

In a broad sense, regionalization is a process by which blood banks within a given geographical area move toward the coordination of their activities. Such coordination may range from cases in which the blood banks merge into a large, centralized unit, to cases where the existing structure remains unaltered and only certain functions, such as donor recruitment, processing and distribution, are coordinated among the blood banks. In most of these cases, questions of optimal region size, central and local bank locations, regional boundaries, optimal distribution and communication network configurations must be answered. Also, administrative policies, ordering and cross-matching policies, and donor recruitment and component therapy strategies must be analyzed and coordinated.

In an earlier paper on regionalization, (Cohen [1975]), hierarchical structures for different types of regional blood banking systems were discussed. The appropriate structure for any area depends on a complex interaction between the level of activity, the economies or diseconomies of scale, the cost effectiveness and efficiency, as well as the interactions of the interested parties. In this section, the impact of size and structure on the different costs of operation of blood banking activities in a region will be analyzed. The main determinants of cost are the personnel, the space, the equipment, the location and allocation of facilities and the transportation used to carry out the activities of the blood center.

Since regions vary in terms of their geography, number of donors, and number of recipients of blood services, the most effective and efficient organizational structure will also vary from region to region. It is not our purpose here to delve into all the many different ramifications of the different regional structures. Rather for this analysis it is sufficient to consider three generic structures of regional blood banking in the United States. Most systems which do not exactly fit into one of these structures can be closely approximated by one of them. The structures are given in Figure 1, and represent (i) a community blood center which services the entire needs of a particular region, (ii) a collection of (communicating) community blood centers each under its own independent control, which services the blood needs of a particular region, and (iii) a regional blood center which either coordinates or controls the activities of a collection of community blood centers which in turn fulfill the needs of the community in the region. As a general rule, as the size of a region changes due to an increase in demand or

CENTRALIZATION ⟷ COORDINATION MODELS

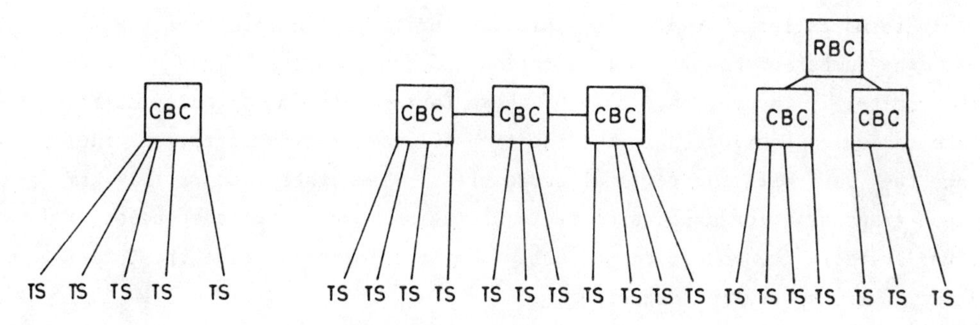

RBC ≡ REGIONAL BLOOD CENTER

CBC ≡ COMMUNITY BLOOD CENTER

TS ≡ TRANSFUSION SERVICE

Centralization may involve:

- Power Transfer
- Centralized Information System
- Mergers, Relocations

Coordination may involve:

- Cooperative Donor Recruitment
- Resource Sharing
- Distribution and Processing
- Share Information
- Coordinate Planning

Figure 1: Regional Blood Banking

geography, the single community blood center may not adequately fulfill the needs of the region and one of the other two structures tends to replace the single center over time. In some cases, the new structure consists of a regional community blood center with satellite facilities.

In the next section, we focus on the location-allocation-transportation aspects of regionalization. Our variables will be central bank locations, regional boundaries and blood distribution network configurations. In the first model, all the other aspects of regionalization are summarized by certain terms called system costs, which are primarily functions of two factors: the number of hospitals in a region and the amount of blood used by each hospital in the region. Both of these factors are functionally related to the variables considered in the location-allocation-transportation model (since they vary with the regional boundaries.) The other factors that affect the non-transportation aspects of regionalization tend to be independent of the variables in the model, consequently they are independent of the location-allocation-transportation decisions.

The third section describes some middle level and institutional level models in their abstract forms. Although these mathematical models are directly applicable to the blood bank decision process, they do provide qualitative insights into the considerations underlying the optimal decisions and the general structure of such decisions.

The final section briefly summarizes the findings.

2. THE REGIONAL MODEL AND DECISIONS

The regionalization model is verbally described as follows: "Within a given geographical area, there are N hospitals. Regionalization is to be achieved by dividing the area into M regions and establishing a central blood bank in each region. All blood banking activities in a region are to be coordinated. Supply generation is to be done mainly by each central bank and each hospital is to obtain its primary blood supply from the central bank in its region. The blood distribution operation consists of periodic and emergency deliveries. The hospitals in a region receive their periodic daily requirements from their central bank. The blood deliveries are made by vehicles which, starting from the central bank, visit one by one the hospitals they are scheduled to supply, and return to the central bank. These vehicles have given capacities and given limits on the number of deliveries they can make per day. Because of the wide fluctuations in demand, a hospital may deplete certain blood types before the next periodic delivery is due. In that case, a delivery vehicle is dispatched immediately, from its central bank.

The delivery vehicle makes an emergency blood delivery to that hospital and returns to the blood bank. The problem is to decide how many central blood banks to set up, where to locate them, how to allocate the hospitals to the banks, and how to route the periodic supply operation, so that the total of transportation costs (periodic and emergency supply costs) and the other system costs are a minimum." This problem will be called the Blood Transportation-Allocation Problem (BTAP).

In modeling the BTAP it is necessary to describe the costs as functions of the decision variables. The periodic delivery costs, which are a set of linear terms, depend on all three types of variables in this problem, that is, routing the periodic supply operations, locating the banks, and allocating the hospitals to the banks. The emergency costs are also a set of linear terms and depend on only two types of variables, locating the banks and allocating the hospitals. The system costs are nonlinear functions of the size of the blood banks and the number of hospitals allocated to the blood banks, and therefore, of all the variables in this problem they depend only on hospital allocations. So, if the system costs were constant and the emergency costs were negligible, the model would be equivalent to the General Transportation Problem (GTP) (see Or [1976], or Magnanti et al. [1975]). If the system costs were constant and the periodic delivery costs were negligible, the model would reduce to a Location-Allocation problem (LAP) (see Cooper [1963, 1964], Hurter and Wendell [1973a, 1973b], or Francis and White [1974]). So the model is a complex combination of these two large problems. The basic strategy we use in order to obtain a good solution depends heavily on these two subproblems. We solve each subproblem independently and then combine them at the end, making tradeoffs between them and superimposing the system costs considerations, to obtain a good solution to the model. However, it should be noted that, unlike the GTP, solving the LAP does not produce a complete, feasible solution to the main model. It only gives the locations and the allocations, and in order to get the missing periodic delivery routes, one must solve a set of vehicle dispatch problems.

Other work on regionalization or centralization of blood bank activities does not consider the location-allocation-transportation aspects. Jennings [1970, 1972] used a simulation model to construct part of a regional blood banking system. He grouped a number of identical hospitals together; however, he did not have a central blood bank. Transshipment policies and inventory levels were studied to see their impact on shortages and outdates. Yen [1965] also studied multiechelon inventory systems. He concentrated his efforts on the optimal inventory levels and the optimal issuing policies. Prastacos [1977] and Prastacos et al. [1977] were interested in the allocation of existing stocks among the hospitals. Neither Jennings, Yen nor Prastacos et al. studied the location of central banks or the allocation of hospitals to them.

THE BLOOD TRANSPORTATION-ALLOCATION MODEL

Because the BTAP is a complex optimization problem, we will make a few reasonable assumptions to decompose the problem into smaller subproblems.

ASSUMPTION 1: The number of banks, M, is a given number.

Even in cases in which the above assumption doesn't hold, M is almost always restricted to a small, finite, feasible set (M is always an integer and $1 \leq M \leq N$). So, in those cases one could solve the problem for each feasible value of M to get the optimal solution. In this respect, *Assumption 1* is not restrictive.

ASSUMPTION 2: The blood delivery period is daily for each hospital.

Considering that some hospitals use more than 7000 units of blood per year, while some others use less than 10, this is an unrealistic assumption. Unfortunately, determining the optimal multiple delivery periods as well as the location-allocation and routings increases the complexity of the problem considerably and makes it almost impossible to find a direct solution procedure. A simple, multiple period problem should have three options for the periodic deliveries (daily, biweekly, weekly). However, the problem of choosing optimal periods for each hospital would be a very large (and time consuming) combinatoric process. If the periods are set in advance (daily, biweekly, weekly) these multiple periods can be easily incorporated into the present location-allocation model merely by adjusting the costs to reflect costs per day. The routes of the delivery vehicles however would have to be adjusted later.

ASSUMPTION 3: The potential locations of the M banks are a finite (and usually small) number.

The set of practical feasible locations is almost always a small, finite set (usually one does not want to build a blood bank from scratch; instead, they are most frequently located in the area's largest hospitals or at existing blood centers). So, if necessary, we could solve the problem for all combinations of feasible locations, to get the optimal solution. In a design problem, this is not an impossible enumeration, since even a region the size of the Chicago metropolitan area (using the amount of blood transfused as a measure of hospital size) has only six central blood banks and has only seven hospitals with consumption rates of over 7000 units per year.

Finally, it is assumed that each vehicle makes one (non-emergency) trip per day. If multiple trips per day are allowed in a real setting this model would need to be modified appropriately. Most of the latter do not qualify to be central blood banks for various reasons. So, in this respect *Assumption 1* is not very restrictive.

The following notation will be used in formulating the BTAP.

i) N is the number of demand points (i.e., hospitals).

ii) M is the number of supply points (i.e., banks)

iii) n is the maximum number of supply vehicles available.

iv) $\mathcal{H} = \{H_1, \ldots, H_N\}$ is a set of N demand points.

v) $\mathcal{S} = \{H_{N+1}, \ldots, H_{N+M}\}$ is a set of M supply points.

vi) $\mathcal{N} = \mathcal{H} \cup S$ is the set of all points involved in the problem.

vii) d_{ij} is the "distance" from H_i to H_j. It should be noted that although Euclidean distances among locations of hospitals and central banks are used in the solution procedure, one could obtain a matrix of accurate travel times between all pairs of hospitals and banks, and one could use this matrix or any other "distance measure" instead of the Euclidean distance matrix.

viii) C_k, $k = 1, \ldots, n$ is the capacity of supply vehicle k.

ix) Q_i, $i = 1, \ldots, N$ is the requirement of demand point i.

x) D_k, $k = 1, \ldots, n$ is the maximum distance supply vehicle k may travel on a non-emergency delivery route.

xi) γ_i, $i = 1, \ldots, N$ is the expected number of emergency deliveries to hospital H_i per period. γ_i is the probability that the demand at H_i exceeds the supply at H_i given the optimal inventory level at H_i is used.

xii) $s(\ell, q)$ is the systems cost function of a region, where ℓ is the number of hospitals in that region, and q is the amount of blood used per year in that region.

xiii) y_{ij}, $i = 1,\dots,N$; $j = N+1,\dots,N+M$ is a zero-one variable such that y_{ij} is 1 if hospital H_i is assigned to central bank H_j and is 0 otherwise.

xiv) x_{ijk}, $i = 1,\dots,N+M$; $j = 1,\dots,N+M$; $k=1,\dots,n$ is a zero-one variable such that x_{ijk} is 1 if vehicle k goes from hospital H_i to H_j and is 0 otherwise.

The BTAP is:

PROBLEM 1:

$$\min z^1(x,y) = \sum_{i=1}^{N+M} \sum_{j=1}^{N+M} \sum_{k=1}^{n} d_{ij} x_{ijk} + \sum_{i=1}^{N} \sum_{j=N+1}^{N+M} \gamma_i d_{ij} y_{ij}$$

$$+ \sum_{j=N+1}^{N+M} s\left(\sum_{i=1}^{N} y_{ij}, \sum_{i=1}^{N} Q_i y_{ij}\right) \qquad (1)$$

subject to

$$\sum_{k=1}^{n} \sum_{j=1}^{N+M} x_{ijk} = 1 \qquad i = 1,\dots,N \qquad (2)$$

$$\sum_{j=1}^{N+M} \sum_{i=1}^{N} Q_i x_{ijk} \le C_k \qquad k = 1,\dots,n \qquad (3)$$

$$\sum_{j=1}^{N+M} \sum_{i=1}^{N+M} d_{ij} x_{ijk} \le D_k \qquad k = 1,\dots,n \qquad (4)$$

$$\sum_{\{i:H_i \in S\}} \sum_{\{j:H_j \in \bar{S}\}} \sum_{k=1}^{n} x_{ijk} \ge 1 \qquad \text{for all } (S, \bar{S}) \qquad (5)$$

where S is any proper subset of $\mathcal{H}$ containing $\mathcal{S}$ and $\bar{S}$ is the complement of S.

$$\sum_{j=1}^{N+M} x_{hjk} = \sum_{i=1}^{N+M} x_{ihk} \qquad k = 1,\dots,n;\ h = 1,\dots N+M \qquad (6)$$

$$y_{ij} \geq \sum_{h=1}^{N+M} x_{ihk} + \sum_{h=1}^{N+M} x_{jhk} - 1 \qquad i = 1,\dots,N;\ j = N+1,\dots,N+M;\ k = 1,\dots,n \tag{7}$$

$$x_{ijk} = 0,1 \qquad i = 1,\dots,N+M;\ j = 1,\dots,N+M;\ k = 1,\dots,n \tag{8}$$

(note that $x_{iik} = 0$)

$$y_{ij} = 0,1 \qquad i = 1,\dots,N;\ j = N+1,\dots,N+M. \tag{9}$$

The explanation of these constraint sets are as follows. Constraints (2) require that every hospital receive a shipment from some vehicle; (3) are the vehicle capacity constraints; (4) are the maximum travel distance constraints (note, it is implicitly assumed that $Q_i \leq C_k$ for $i=1,\dots,N$ and $k=1,\dots,n$); (5) require that graph $\mathcal{L}$ corresponding to x is connected; (6) imply that a vehicle departs from a point h if and only if it enters there (conservation of flow); (7) contains the coupling constraints between variables $x = \{x_{ijk}\}$ and $y = \{y_{ij}\}$. It means that if there is vehicle k passing from both hospital i, $(\sum_{h=1}^{N+M} x_{ihk} = 1)$, and from bank j, $(\sum_{h=1}^{N+M} x_{jhk} = 1)$, then hospital i is assigned to bank j, $(y_{ij} \geq 1+1-1 = 1)$. As shown in Or [1976], these constraints imply that there is an optimal solution in which each vehicle is based at a particular supply point.

In *Problem 1* the variables $x = \{x_{ijk}\}$ correspond to the routing of the periodic delivery vehicles and the variables $y = \{y_{ij}\}$ correspond to the allocations of the hospitals to the blood banks. For a given $x = \{x_{ijk}\}$, $y = \{y_{ij}\}$ is uniquely determined, but the converse is not true; if we are given the allocations, a series of M vehicle dispatch problems have to be solved, in order to obtain the routings. *Problem 1* has a finite feasible solution set and a nonempty optimal solution set. However, the underlying Multiple Vehicle Dispatch Problem (MVDP) makes it a complex integer programming problem. For N of any significant size ($N \geq 20$), the BTAP is too large to be solved by conventional mathematical programming techniques in a reasonable amount of time.

In *Problem 1,* if γ_i, $i = 1,\dots,N$ are small or emergency costs negligible (actual γ_i's range from .0002 to .06 when optimal ordering policies are followed, see Pierskalla and Yen [35]) and the function $s(\ell, k)$ is essentially constant, then

$$z^2(x) = \sum_{i=1}^{N+M} \sum_{j=1}^{N+M} \sum_{k=1}^{n} d_{ij} x_{ijk}$$

would be the dominating term in the objective function (1). Then we could just solve the MVDP,

<u>*PROBLEM 2*</u>

$$\min \sum_{i=1}^{N+M} \sum_{j=1}^{N+M} \sum_{k=1}^{n} d_{ij} x_{ijk} \qquad (11)$$

subject to

$$\sum_{k=1}^{n} \sum_{j=1}^{N+M} x_{ijk} = 1 \qquad i=1,\dots,N \qquad (12)$$

$$\sum_{j=1}^{N+M} \sum_{i=1}^{N} Q_i x_{ijk} \leq C_k \qquad k=1,\dots,n \qquad (13)$$

$$\sum_{j=1}^{N+M} \sum_{i=1}^{N+M} d_{ij} x_{ijk} \leq D_k \qquad k=1,\dots,n \qquad (14)$$

$$\sum_{\{i:H_i \in S\}} \sum_{\{j:H_j \in \bar{S}\}} \sum_{k=1}^{n} x_{ijk} \geq 1 \qquad \text{for all } (S, \bar{S}) \qquad (15)$$

$$\sum_{j=1}^{N+M} x_{hjk} = \sum_{i=1}^{N+M} x_{ihk} \qquad k=1,\dots,n;\ h=1,\dots,N+M \qquad (16)$$

$$x_{ijk} = 0, 1 \qquad i=1,\dots,N+M;\ j=1,\dots,N+M;\ k=1,\dots,n \qquad (17)$$

in order to obtain the optimal x^* for <u>*Problem 1.*</u> The optimal allocations, y^*, would then be uniquely determined by x^*.

On the other hand, if γ_i, i=1,...,N are relatively large (which might happen under nonoptimal ordering policies) or system costs and periodic delivery costs are negligible, then

$$z^3(y) = \sum_{i=1}^{N} \sum_{j=N+1}^{N+M} \gamma_i d_{ij} y_{ij}$$

would be the dominating term in the objective function. Then we could just solve the allocation problem,

PROBLEM 3

$$\min \sum_{i=1}^{N} \sum_{j=N+1}^{N+M} \gamma_i d_{ij} y_{ij} \tag{18}$$

subject to

$$\sum_{j=N+1}^{N+M} y_{ij} = 1 \qquad i = 1,\ldots,N \tag{19}$$

$$y_{ij} = 0,\ 1 \qquad i = 1,\ldots,N;\ j = N+1,\ldots,N+M \tag{20}$$

in order to get the optimal y° for *Problem 1*. Then, optimal routings, x°, would be obtained by solving a vehicle dispatch problem for each one of the M regions determined by y°.

Let x^* be an optimal solution of *Problem 2*. Let y^* be the allocations determined by x^*. Let y° be an optimal solution of *Problem 3*.

It directly follows from the above definitions that

$$z^2(x^*) \leq z^2(x^\circ)$$

$$z^3(y^\circ) \leq z^3(y^*)$$

and if the systems costs are essentially constant, then

$$z^2(x^*) + z^3(y^\circ)$$

would be a good lower bound on the optimal value of *Problem 1*.

In order to minimize the cost of operation under different regional system configurations, the economies of scale curves representing feasible combinations of the functional areas of blood banking were incorporated into the model. These curves form the basis for determining the operating costs for each of the different regional structures.

For example, if one considers the regional structure given by the first figure in Figure 1, then the only questions which arise relative to costs involve where should the community blood center be located in order to minimize total transportation costs for donors, recruiting, phlebotomy[3] on mobiles, and routine and emergency deliveries to the transfusion services. Since the other costs, such as processing, administration, inventory control and phelbotomy at the Center are relatively independent of location, these costs would not be included in the decision process, because they would be incurred no matter where the community blood center was located.

On the other hand, if the regional structure is that of the second figure in Figure 1, then all of the costs from the economies of scale curves are relevant, since not only the location of the community blood centers but also their sizes in the region are important decision variables. Consequently, all of the curves are needed, and tradeoffs among these costs and locations in sizes must be made.

Finally, if one were to analyze the third figure of Figure 1, the appropriate use of the economies of scale curves would depend upon the authority structure and governance relationships between the regional blood center and the community blood centers. For example, if the regional blood center were only a coordinating body of information and did not really have any authority over the community blood centers, then the costs of operation would be very similar to those in the preceding paragraph. That is, all of the functional areas of blood banking would be located at the community blood centers; consequently, virtually all of the costs would be incurred there, and hence if one were to do a regional design, one would be interested in the location of the blood centers, as well as their operating size. On the other hand, if for example, donor recruiting were done at the regional blood center, then the costs of donor recruiting would not be charged to the community blood centers but would be a regional blood center cost. Since donor recruiting costs are related to the distances the donor recruiters have to travel, then the location of the regional blood center would be a factor in the cost structure; however, none of the other costs, such as phlebotomy, processing, inventory control and distribution administration at the community blood

---

[3]Phlebotomy is the procedure of drawing blood from a donor. These drawings often occur at the hospital, blood center, or at a distant location such as a school, church, company, etc. When drawings are made at a distant location, mobile blood vehicles are used and the drawings are thus called "phlebotomy on mobiles."

centers would be included in the cost of the regional blood center itself. Consequently, wherever the functional areas and authority for administration of those areas are located in the system, then those costs should be charged at those appropriate places.

3. DECISIONS AT LOWER ECHELONS

The preceding analysis focused on macro and some middle level decision making models in a regional system. In this section, we discuss decisions on inventory control, issuing, and crossmatch release policies at lower levels in the modeling hierarchy. These decisions impact the decisions at the higher levels since they affect the system cost structures mentioned in the previous section.

The basic community blood center scheme is that of a "wheel" structure as shown in Figure 2, which is another view of Figure 1a.

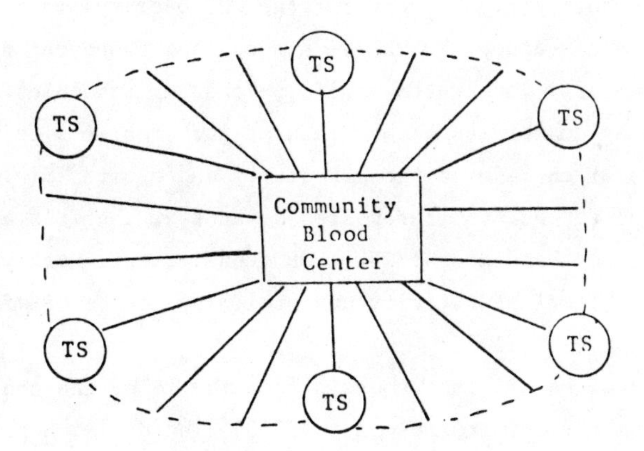

Figure 2

The Community Blood Center and Its
Satellite Transfusion Service (TS's)

The community blood center acts as the hub of activities for its satellite blood banks and transfusion services. In the centralized system, the central blood bank supplies a number of TS's and maintains the authority to redistribute all blood in the system. Most TS's maintain a supply of whole blood and/or packed red cells. Other components may also be maintained at some facilities.

Another way to view a CBB and the decisions needed to answer the questions posed earlier is shown in Figure 3. In this figure, the CBB is shown at the top and the lines represent the flow of units through the different activities and TS's in the system. That is, Figure 3 is a schematic drawing of the inputs, outputs, and flows in a centralized system. Some of the decisions needed by the CBB are shown in the diamond-shaped boxes. The variables $S_0, S_1, \ldots, S_N$ represent the number of units needed on hand each day at the CBB and at the TS's in order to meet the system needs without excess outdates. Of course, the S_i's are different for each ABO-Rh type and each component[4] and they change over time depending upon the changing needs at the TS's.

Basically the inventory flow system for the CBB operates in the following manner. Forecasts of future ABO-Rh blood needs and component needs are made. The CBB periodically constructs mobile phlebotomy schedules and forecasts the corresponding quantities to be drawn at each mobile site. Individual drawings are also often made at the CBB itself and/or its TS's. These drawings are scheduled to meet the forecasted demands at the TS's and maintain a stock of inventories on hand at the CBB. On a daily, semi-weekly or weekly basis depending on the level of activity and proximity to the CBB for each TS, orders to the TS's must be filled.

After the CBB receives all the requests from the TS's, the orders are filled by drawing from the inventories in the CBB. The decision policy as to which units to send is called the issuing policy. The most common issuing policies are first-in-first out (FIFO) or last-in-first-out (LIFO). For purposes of simplification as well as good medical practice, each ABO type and Rh factor is considered independent of the other types and Rh factors. When the sum of all TS demands for whole blood/packed red cells (WB/PRC's) or for particular components exceeds the total inventory in the CBB, the CBB may backlog the excess demand or may fill all demands by calling in donors, by contacting other CBB's, by using frozen packed red cells if appropriate, or

---

[4]There are many derivative products which comprise whole blood and which can be separated from the whole blood. These derivatives are called components. Some commonly derived components are: platelets, granulocytes, leukocyles, cryoprecipitate, plasma and red cells.

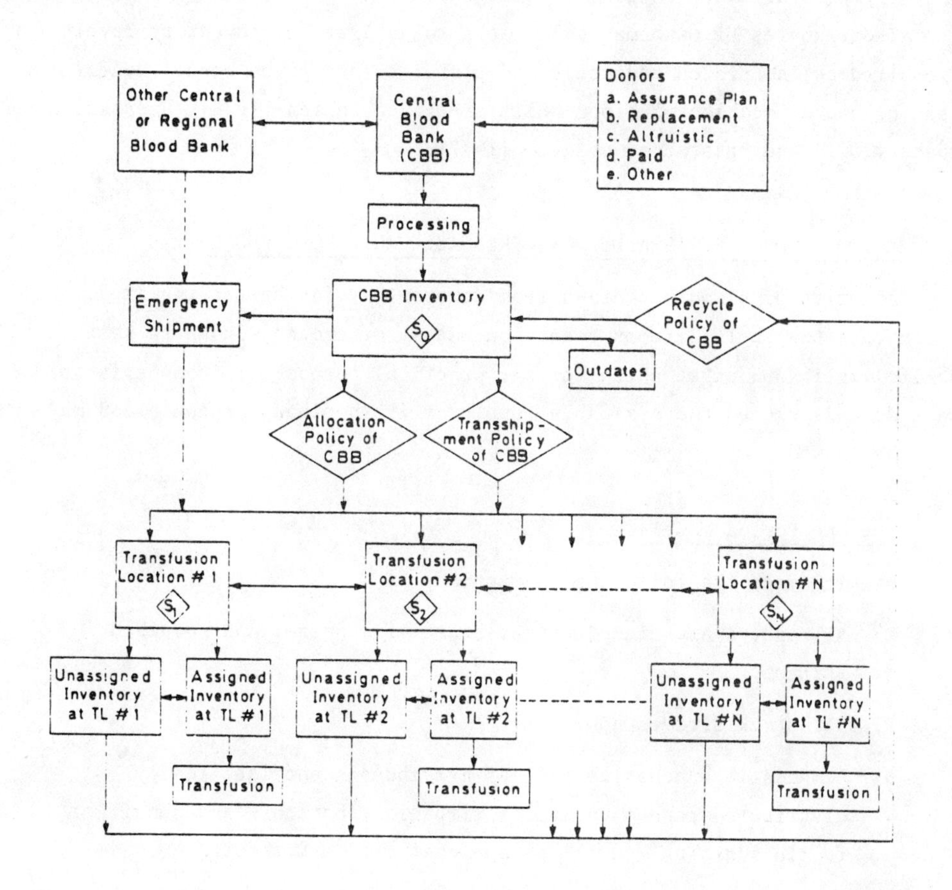

FIGURE 3.

FLOW CHART FOR A CENTRAL BLOOD BANKING SYSTEM SHOWING INVENTORY LOCATIONS AND CBB DECISION POLICIES

by requesting an emergency shipment from still higher echelon blood banks. The CBB uses different approaches to handle the excess demand depending upon whether the orders are routine or emergency.

Two examples of mathematical models to analyze the inventory levels for fresh red cells, frozen red cells and platelets are given here. Models for issuing and allocation of blood units are given in Yen [1975], Pierskalla and Roach [1972] and Prastacos, et al., [1977].

4. A TWO PRODUCT PERISHABLE/NON PERISHABLE INVENTORY PROBLEM

As units of blood are drawn from donors they may be refrigerated, in which case the shelf life is twenty-one days, or frozen, in which case the shelf life is 365 days, which for all practical purposes is non-perishable. When demands exceed the available supply of fresh blood, frozen blood may then be thawed.

The One Period Model

We will make the following assumptions:

1) All orders are placed at the start of a period and received instantly.
2) All stock arrives new.
3) Demands in successive periods are independent identically distributed random variables with distribution F and density f. In addition we will assume that $f(t) > 0$ when $t > 0$.
4) Inventory of product 1 (the perishable product) is depleted according to a FIFO policy.
5) All costs are linear. They include:
   a) Ordering in both products (at unit costs c_1 and c_2) charged at the start of the period;
   b) Holding in both products (at unit costs h_1 and h_2) charged on what is on hand at the end of the period;
   c) Shortage (at unit cost r) charged on the unsatisfied demand at the end of the period (the shortage cost is either the cost of emergency shipments from some source on the marginal cost of an additional day in the hospital for postponed surgery);
   d) Outdating (at unit cost θ) charged on what deteriorates at the end of the period.
6) If product 1 has not been depleted by demand before reaching

age m-periods, it must be discarded at the unit cost given in 5(d).

7) There is a single demand source. Demands first deplete from product 1 (the perishable product) and then product 2. Excess demand is backlogged in the second product.

A number of the assumptions may be relaxed or altered. It should be noted that assumption (4) is quite mild. The optimality of FIFO for depletion of perishable inventory has been established under far more general circumstances (Nahmias [1974] and Pierskalla and Roach [1972]).

The approach will be to charge the outdating cost against the expected outdating of the present order which will not occur for m periods. The motivation behind this method is discussed in Nahmias [1975]. If $\underset{\sim}{x} = (x_{m-1},\ldots,x_1)$ is the vector of perishable inventory on hand, x_i = number of units on hand which will outdate in exactly i periods, and y is the amount of new perishable inventory ordered, then it has been shown that the expression $\int_0^y G_m(u; \underset{\sim}{x})du$ represents the expected outdating of y, m periods into the future, where

$$G_n(t; \underset{\sim}{x}(n-1)) = \int_0^t G_{n-1}(x_{n-1} + v; \underset{\sim}{x}(n-2))f(t-v)dv$$

for $1 \leq n \leq m$; $\underset{\sim}{x}(n) = (x_n,\ldots,x_1)$ and $G_0(t) = \begin{cases} 1 & \text{if } t \geq 0 \\ 0 & \text{if } t < 0 \end{cases}$. For each $n \geq 1$, $G_n(t; \underset{\sim}{x}(n-1))$ is a C.D.F. in its first argument, and may possess a discontinuity at t = 0.

Letting x^2 = amount of product 2 on hand,

z = amount of product 2 on hand after ordering,

the total expected cost of ordering y of product 1 and $z - x^2$ of product 2 is

$$c_1y + h_1 \int_0^{x+y} (x+y-t)f(t)dt + \theta \int_0^y G_m(u; \underset{\sim}{x})du + c_2(z-x^2) + h_2zF(x+y)$$

$$+ h_2 \int_{x+y}^{x+y+z} (x+y+z-t)f(t)dt + r \int_{x+y+z}^{\infty} (t-x-y-z)f(t)dt$$

where $x = \sum_{i=1}^{m-1} x_i$ for convenience. Collecting all terms independent of x^2 we will write this as $L(\underset{\sim}{x}, y, z) - c_2x^2$. A point which should be noted here is that the decision variables for each product have different interpretations: y represents the actual quantity of product 1 ordered, while z is the inventory level of product 2 after ordering. Our interest in the single period model is secondary to that of the multi-decision dynamic problem. The optimal ordering policy over the finite horizon will satisfy the functional equations

$$C_n(\underset{\sim}{x}, x^2) = \inf_{\substack{y\geq 0 \\ z\geq x^2}} \{L(\underset{\sim}{x}, y, z) - c_2x^2 + \alpha\int_0^\infty C_{n-1}(\underset{\sim}{s}_1(\underset{\sim}{x}, y, t), s_2(x+y, z, t))f(t)dt\}$$

for $1 \leq n \leq N$(N is a fixed positive integer). The transfer functions are given by:

$$\underset{\sim}{s}_1(\underset{\sim}{x}, y, t) = (s_{1,m-1}(\underset{\sim}{x}, y, t), \ldots, s_{1,1}(\underset{\sim}{x}, y, t))$$

where

$$s_{1,i}(\underset{\sim}{x}, y, t) = (x_{i+1} - (t - \sum_{j=1}^{i} x_j)^+)^+, \quad 1 \leq i \leq m-1$$

$$(\text{we interpret } x_m = y)$$

and $s_2(x+y, z, t) = z - (t-(x+y))^+$ where $g^+ = \max(g, 0)$. The discount factor is $\alpha \in (0, 1)$.

Again collecting all terms independent of x^2, we will write

$$C_n(x, x^2) = \inf_{\substack{y\geq 0 \\ z\geq x^2}} \{B_n(\underset{\sim}{x}, y, z) - c_2x^2\}$$

in which

$$B_n(\underset{\sim}{x}, y, z) = L(\underset{\sim}{x}, y, z) + \alpha \int_0^\infty C_{n-1}(\underset{\sim}{s}_1(\underset{\sim}{x}, y, t), s_2(x+y, z, t))f(t)dt.$$

The functions $C_n(\underset{\sim}{x}, x^2)$ have the usual interpretation as the minimum expected discounted cost for an n period problem when $(\underset{\sim}{x}, x^2)$ is on hand. As is customary with dynamic programming models, the periods are numbered backwards. The goal of our analysis will be to answer the two questions: when should an order be placed and how much of each product should be ordered?

The Ordering Regions

It becomes notationally and mathematically convenient to introduce the assumption that inventory remaining at the end of the horizon can be salvaged; inventory of product 1 remaining may be salvaged at a return c_1x and of product 2 at a return c_2x^2. Backlogged demand in product 2 may be made up by an emergency order at a cost $- c_2x^2$. That is,

$$C_0(x, x^2) = - c_1x - c_2x^2.$$

The assumption is identical to that made by Veinott [1965] in the analysis of non-perishable multi-product problems. The convenience of this assumption is that the somewhat artificial fact that the problem terminates in a given time horizon with inventory remaining is removed and the problem behaves more like a problem under steady state conditions. In some inventory problems this assumption allows the decomposition of the n-period problem into n one-period problems which are easier to solve.

In addition we will make the following four assumptions regarding the cost parameters:

i) $0 \leq h_2 \leq h_1$

ii) $0 < c_1 < c_2$

iii) $r > (1-\alpha)c_2$

iv) $0 \leq (1-\alpha)(c_2 - c_1) + (h_2 - h_1) < \theta$

Since perishable inventory must often be stored under special conditions, assumption (i) is not unreasonable. Assumption (ii) is necessary to insure that it is economical to stock product 1. Assumption (iii) is a usual one made for non-perishable inventory (Arrow, et al., [1958]). The expression of assumption (iv) is precisely the difference in the costs of procuring and holding one unit of each type of inventory and salvaging it the following period. If this term were negative then it would never be optimal to order in product 1 ; or if it exceeded the unit cost of outdating, it would never be optimal to order to a positive level in product 2.

We will have need to refer to the following constants in the analysis of the ordering regions in the first period:

$$u^* = F^{-1}\left[\frac{r-c_2(1-\alpha)}{r+h_2}\right]$$

$$w^* = F^{-1}\left[\frac{r-c_1(1-\alpha) + (h_2 - h_1)}{r+h_2}\right].$$

When F is strictly increasing, assumptions (iii) and (iv) guarantee that u^* and w^* both exist and are strictly positive. If the additional condition $\alpha c_2 - h_2 < c_1$ is satisfied (although we will not require it to be), then we may also define the constant

$$v^* = F^{-1}\left[\frac{r - c_1 + \alpha c_2}{r+h_2}\right]$$

With the inclusion of the salvage value assumption it follows from the definition of the transfer functions that

$$\begin{aligned} B_1(\underset{\sim}{x}, y, z) = {} & L(\underset{\sim}{x}, y, z) + \alpha F(x_1)[- c_1(y + \sum_{i=2}^{m-1} x_i) - c_2 z] \\ & - \alpha c_1 \int_{x_1}^{x+y} (x+y-t)f(t)dt - \alpha c_2[F_2(x+y) - F(x_1)]z \\ & - \alpha c_2 \int_{x+y}^{\infty} (x+y+z-t)f(t)dt) \end{aligned}$$

We will adopt the following notational convention: if $h: R^n \to R^1$ and $h \in C^{(2)}$, then $h^{(i)}$ is the first partial derivative of h with respect to its ith argument and $h^{(i,j)}$ is the second cross partial derivative with respect to the $i^{\underline{th}}$ and $j^{\underline{th}}$ arguments respectively.

We have the following

THEOREM II.4.1: $B_1(\underset{\sim}{x}, y, z)$ is convex in (y, z) for all non-negative $\underset{\sim}{x}$. The functions $y_1(\underset{\sim}{x})$, $z_1(\underset{\sim}{x})$ solving $B_1(\underset{\sim}{x}, y_1(\underset{\sim}{x}), z_1(\underset{\sim}{x})) = \min_{y,z}\{B_1(\underset{\sim}{x}, y, z)\}$ satisfy $B_1^{(m)}(\underset{\sim}{x}, y_1(\underset{\sim}{x}), z_1(\underset{\sim}{x})) = B_1^{(m+1)}(\underset{\sim}{x}, y_1(\underset{\sim}{x}), z_1(\underset{\sim}{x})) = 0$ and are unique for all $\underset{\sim}{x}$.

Since $y_1(\underset{\sim}{x}) > 0$ for all $\underset{\sim}{x}$, a necessary and sufficient condition that it be possible to order to the global minimum of $B_1(\underset{\sim}{x}, y, z)$ is $x^2 < z_1(\underset{\sim}{x})$. If $x^2 \geq z_1(\underset{\sim}{x})$, it may still be optimal to order product 1. In this case we define the function $p_1(\underset{\sim}{x}, x^2)$ to satisfy

$$B_1(\underset{\sim}{x}, p_1(\underset{\sim}{x}, x^2), x^2) \equiv \inf_y(B_1(\underset{\sim}{x}, y, x^2)).$$

If $p_1(\underset{\sim}{x}, x^2) > 0$ then it is optimal to order this amount of product 1.

Hence the following characterization follows:

THEOREM II.4.2: A necessary and sufficient condition that it is optimal to order a positive amount of product 1 is $B_1^{(m)}(\underset{\sim}{x}, 0, x^2) < 0$. If one reasons analoguously to *THEOREM II.4.2*, then it is tempting to assume that $B_1^{(m+1)}(\underset{\sim}{x}, 0, x^2) < 0$ implies it is optimal to order in product 2. However this is not the case. To see why, let $t(\underset{\sim}{x}, y)$ satisfy $B_1^{(m+1)}(\underset{\sim}{x}, y, t(\underset{\sim}{x}, y)) = 0$. Differentiating implicitly with respect to y we obtain

$$t^{(m)}(\underset{\sim}{x}, y) = \frac{-B_1^{(m+1,\, m)}(\underset{\sim}{x}, y, t(\underset{\sim}{x}, y))}{B_1^{(m+1,\, m+1)}(\underset{\sim}{x}, y, t(\underset{\sim}{x}, y))} < 0.$$

Since $t(\underset{\sim}{x}, y_1(\underset{\sim}{x})) = z_1(\underset{\sim}{x})$ and $y_1(\underset{\sim}{x}) > 0$ it follows that $t(\underset{\sim}{x}, 0) > z_1(\underset{\sim}{x})$. If x^2 satisfies $z_1(\underset{\sim}{x}) \leq x^2 < t(\underset{\sim}{x}, 0)$ then $B_1^{(m+1)}(\underset{\sim}{x}, 0, x^2) < 0$ and it is optimal not to order in product 2.

Hence there are exactly three distinct ordering regions:

Region I - Optimal to order in both products: $x^2 < z_1(\underset{\sim}{x})$,

Region II - Optimal to order in product 1 only: $x^2 \geq z_1(\underset{\sim}{x})$
$B_1^{(m)}(\underset{\sim}{x}, 0, x^2) < 0(p_1(\underset{\sim}{x}, x^2) \geq 0.$

Region III - Optimal not to order: $B_1^{(m)}(\underset{\sim}{x}, 0, x^2) \geq 0$, $(p_1(\underset{\sim}{x}, x^2) \leq 0$.

Note from the proof of *THEOREM II.4.2* that if it is optimal to order in product 2 then it is optimal to order in product 1. The boundary between Regions I and II is quite complex, as it depends on the entire vector $(\underset{\sim}{x}, x^2)$. However the boundary between Regions II and III depends on the vector $\underset{\sim}{x}$ only through the sum of its components, x.

Define $g(x) = (1/r+h_2)) \cdot \{r + \alpha c_2 - c_1 - F(x)[(h_1 - h_2) + \alpha(c_2 - c_1)]\}$. Then

THEOREM II.4.3: A necessary and sufficient condition that $(\underset{\sim}{x}, x^2)$ is in Regions I or II is that

$$F(x + x^2) < g(x).$$

The function g(x) is strictly decreasing in x with $g(+\infty) = F(w^*)$. If the condition $\alpha c_2 - h_2 < c_1$ is satisfied then $F(w^*) \leq g(x) < 1$ for all $x \geq 0$ and $v^* = F^{-1}(g(0))$ will exist. However, if $\alpha c_2 - h_2 \geq c_1$ then $g(0) \geq 1$. In this case there exists a unique number $p^* \geq 0$ which solves $g(p^*) = 1$. The boundary between Regions II and III may be pictured in the (x, x^2) plane

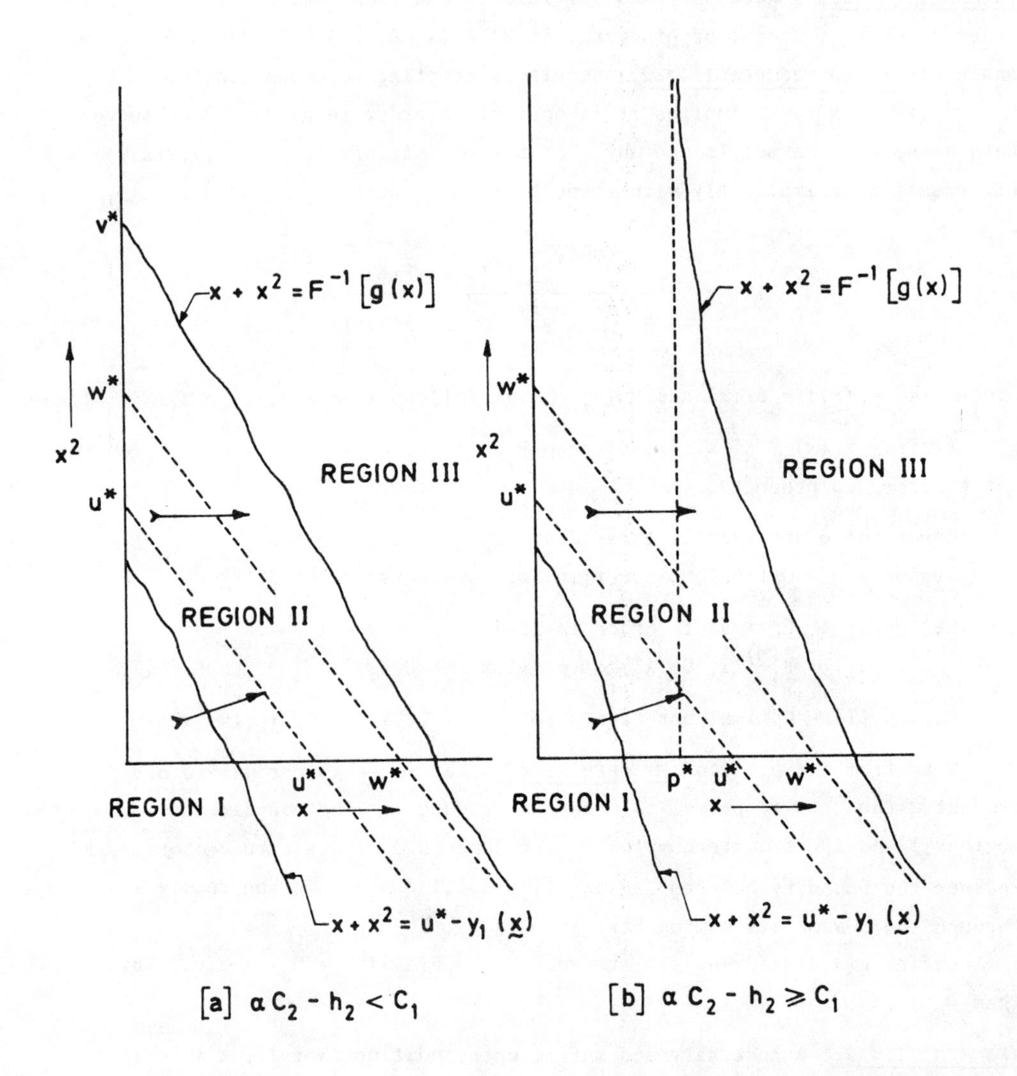

FIGURE 4. THE ORDERING REGIONS

independent of m. Figure 4 pictures this boundary for each of the two cases above. The boundary between Regions I and II may be pictured in the (x, x^2) plane only if m = 2. The arrows indicate the inventory position after ordering. Notice that $(\underset{\sim}{x}, x^2) \in$ Region I guarantees that $x + y_1(\underset{\sim}{x}) + z_1(\underset{\sim}{x}) = u^*$ (which follows from $B_1^{(m+1)}(\underset{\sim}{x}, y_1(\underset{\sim}{x}), z_1(\underset{\sim}{x})) = 0$) while $(x, \tilde{x}^2) \in$ Region II will yield $F(x + x^2 + p_1(\underset{\sim}{x}, \tilde{x}^2)) < g(x + p_1(\underset{\sim}{x}, x^2))$.

These results can be extended to the multiperiod dynamic problem. Even with the inclusion of the salvage value assumption, neither the ordering regions nor the ordering policies are stationary in time.

5. A BY-PRODUCT PRODUCTION SYSTEM WITH AN ALTERNATIVE

The idea behind this example is that components are made from whole blood, e.g., red cells and platelets. In such cases, two simultaneous decisions must be made. First, the inventory level of each product must be determined to best meet (random) demands. Second, the best (optimal) production level (e.g., number of runs) must be determined to achieve these inventories. Often these decisions cannot be made independently because production capacities or design characteristics of the production lines may prohibit arbitrary combinations. The system we consider is describe in Figure 5. There are two inventory items, called products 1 and 2, and two different production processes called types A and B. Type A is capable of making both products, simultaneously, according to the production coefficients η_1 and η_2. When type A is operated at the unit level (e.g., a single run), η_1 and η_2 units of products 1 and 2 are obtained. We choose to call type A a by-product production process due to its multi-product capability. Type B is a single item production process and is capable of making only product 2. In this context, type B is the alternative method of obtaining product 2.

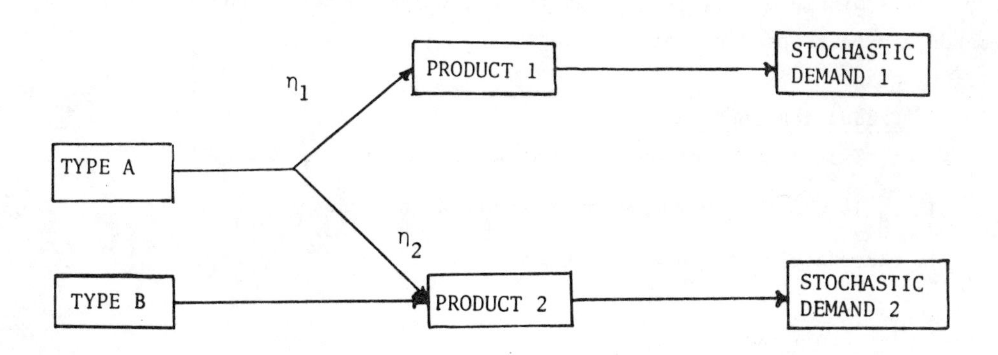

FIGURE 5

Block Diagram of the Production System

The model is of the periodic review type, where the planning horizon is N periods long. At the beginning of period n, n=1,2,...,N, the initial inventories (before production) $x_n = (x_{n,1}, x_{n,2})$ are reviewed, where $x_{n,j}$ is the initial inventory of product j, j=1,2. Then, the starting inventories (stock levels after production but before demand) $y_n = (y_{n,1}, y_{n,2})$ and the the production levels $p_n = (p_{n,A}, p_{n,B})$ are jointly determined. $y_{n,j}$ is the starting inventory of product j, j=1,2 in period n and $p_{n,A}$, $p_{n,B}$ the production levels of types A and B, respectively. Then, a random demand $D_n = (D_{n,1}, D_{n,2})$ is realized. We use a reverse recursion numbering scheme so that index N refers to the first period and 1 refers to the last period. When focusing attention to quantities within a period, we frequently drop the period index on the above quantities.

The following is a list of specific assumptions governing the development of our model.

1) Production Processes

It is assumed that each production process is controlled by specifying the production level. For type A this level is p_A; for type B it is p_B. Thus, the amount of each product produced is $(\eta_1 p_A, p_B + \eta_2 p_A)$. We assume that η_1 and η_2 are constant; they are not decision variables. For the present, we assume that there are no lead times and that there are no upper bounds on production.

2) Demands

We assume that D_n, n=1,2,...,N form a sequence of nonnegative independent and identically distributed random vectors with a continuous joint density function $f(\cdot)$ and finite expected value. Although we allow dependencies among demands within a period, we do require that $D_{n,j}$ be satisfied only from inventories of product j, j=1,2. It is further assumed that unsatisfied demands are backlogged and are not lost.

3) Production Costs

a) c_A is the cost per unit of production on A;

b) c_B is the cost per unit of production on B;

c) there are no fixed costs of production.

4) Inventory Costs

Let $g_i(\cdot)$ be the holding and shortage cost for product i, i=1,2, over any single period. We define the expected total holding and shortage over any period as given by

$$L(y) = \int_0^\infty \int_0^\infty [g_1(y_1 - t) + g_2(y_2 - u)]\, f(t, u)\,dt\, du.$$

We assume that

a) $L(\cdot)$ is strictly convex over R_2;

b) $L(\cdot)$ has continuous second partial derivatives;

c) $\frac{1}{\eta_1}(c_A - \eta_2 c_B)y_1 + c_B y_2 + L(y) \to +\infty$

whenever $||y|| \to +\infty$ (where $||\cdot||$ is the Euclidean norm). (3) assures the existence of an optimal policy.

5) Discount Parameter

Let $\alpha \in [0, 1]$. α denotes the parameter which relates costs in future periods to the present.

We use the following notation for derivatives and partial derivatives of functions. For a twice continuously differentiable scalar function $h(x)$, let $h'(\cdot)$ and $h''(\cdot)$ represent its first and second derivatives, respectively. For a function $g(\cdot)$ defined on R_n with continuous second partial derivatives, let

$$g^{(i)}(x) = \frac{\partial}{\partial x_i} g(x) \quad , \qquad i=1,2,\dots,n$$

$$g^{(i,j)}(x) = \frac{\partial^2}{\partial x_j\, \partial x_i} g(x), \qquad i,j=1,2,\dots,n.$$

The problem is to determine a policy that leads to the minimum expected discounted costs over the N period horizon. This leads to an optimization problem that is most easily conceptualized as a dynamic programming problem. Before formulating the dynamic program, we first show the relationship between a feasible set of production levels $p = (p_A, p_B)$ and starting inventory levels $y = (y_1, y_2)$ given any initial inventory x. Given p and x, the starting inventories y must satisfy

$$y_1 = \eta_1 p_A + x_1$$

$$y_2 = \eta_2 p_A + p_B + x_2.$$

Given y and x, the production levels p are determined by

$$p_A = \frac{1}{\eta_1}(y_1 - x_1)$$

$$p_B = (y_2 - \frac{\eta_2}{\eta_1}y_1) - (x_2 - \frac{\eta_2}{\eta_1}x_1) = y_2 - (x_2 + \frac{\eta_2}{\eta_1}(y_1 - x_1))$$

Although the dynamic program is most easily defined in terms of p anx x, we favor using y and x since they characterize the optimal policy.

The dynamic programming recursion is given by

$$c_n(x) = \inf \{G_n(y) - \frac{1}{\eta_1}(c_A - \eta_2 c_B)x_1 - c_B x_2\}$$

subject to

$$y_1 \geq x_1$$

$$y_2 \geq x_2 + \frac{\eta_2}{\eta_1}(y_1 - x_1)$$

for

$$G_n(y) = \frac{1}{\eta_1}(c_A - \eta_2 c_B)y_1 + c_B y_2 + L(y)$$

$$+ \alpha \int_0^\infty \int_0^\infty C_{n-1}(y_1 - t, y_2 - u)f(t, u)dt\, du$$

where $n=1,2,\ldots,N$, $y \in R_2$ and $C_0(x) = 0$ for all x.

The first result is of a purely technical nature, but provides the properties of $C_n(\cdot)$ and $G_n(\cdot)$ crucial to the characterization theorem.

THEOREM II.5.1: Under the assumptions made above, the following hold.

1. $G_n(\cdot)$ is continuous and strictly convex.
2. All sets of the form $Q_n(\beta) = \{y \in R_2;\ G_n(y) \leq \beta\}$ are compact and convex for each bounded $\beta \in R$.
3. The point $y_n(x)$ that solves the constrained dynamic program exists and is unique. Hence, inf can be replaced by min for each bounded $x \in R_2$ and $n=1,2,\ldots,N$.

4. $C_n(\cdot)$ is convex and nonnegative on R_2.

The optimal policy, $y_n(x)$ is completely characterized by a point $S_n = (S_{n,1}, S_{n,2})$ and three scalar functions $r_n(\cdot)$, $w_n(\cdot)$, and $q_n(\cdot)$ as graphically described in Figure 6. Each arrow represents the path from an initial inventory (the tail) to its beginning inventory (the head) by some production combination. In Region I both A and B are used and notice that the starting inventory level will always be S_n. In Region III only type B is used so the starting stock always lies on the graph of $q_n(\cdot)$. In Region II only A is used. The path from x to $y_n(x)$ in this region is a line parallel to $w_n(\cdot)$. It is also important to point out that $r_n(\cdot)$, $w_n(\cdot)$ and $q_n(\cdot)$ are functions of x_1.

Many of the assumptions in this example can be relaxed or removed so the model is more realistic. The demands and costs may be nonstationary, the production capacities may be finite, unmet demands may be lost, fixed lead times may be incorporated and the general m-process-n-product case follows naturally.

In summary, then the analysis presented in these section allows one to look at the economics of regionalization, including the location of community blood centers, the allocation of hospital blood banks and transfusion services to them, and the costs of different structures for a region. As mentioned at the beginning, these are not the only variables which one would want to consider in implementing a regional system. However, they are extremely important variables since they affect the system operation and its short and long-run costs. These costs in turn affect the subsequent charging mechanisms for blood and components to the patient. This quantitative analysis is a guide to making important decisions in a more enlightened manner.

6) Concluding Remarks

Finally, it is hoped that the preceding analysis has communicated to the reader some of the flavor of past and current research in several pertinent methodological areas of Operations Research: Inventory Theory, Facility Location, Vehicle Scheduling, and to a small extent Mathematical Programming and Optimization of Stochastic Models. In addition it is hoped the reader will be interested in pursuing the design and analysis of models which better measure health care delivery phenomena and which lead to better decisions for individuals and institutions in this important societal area.

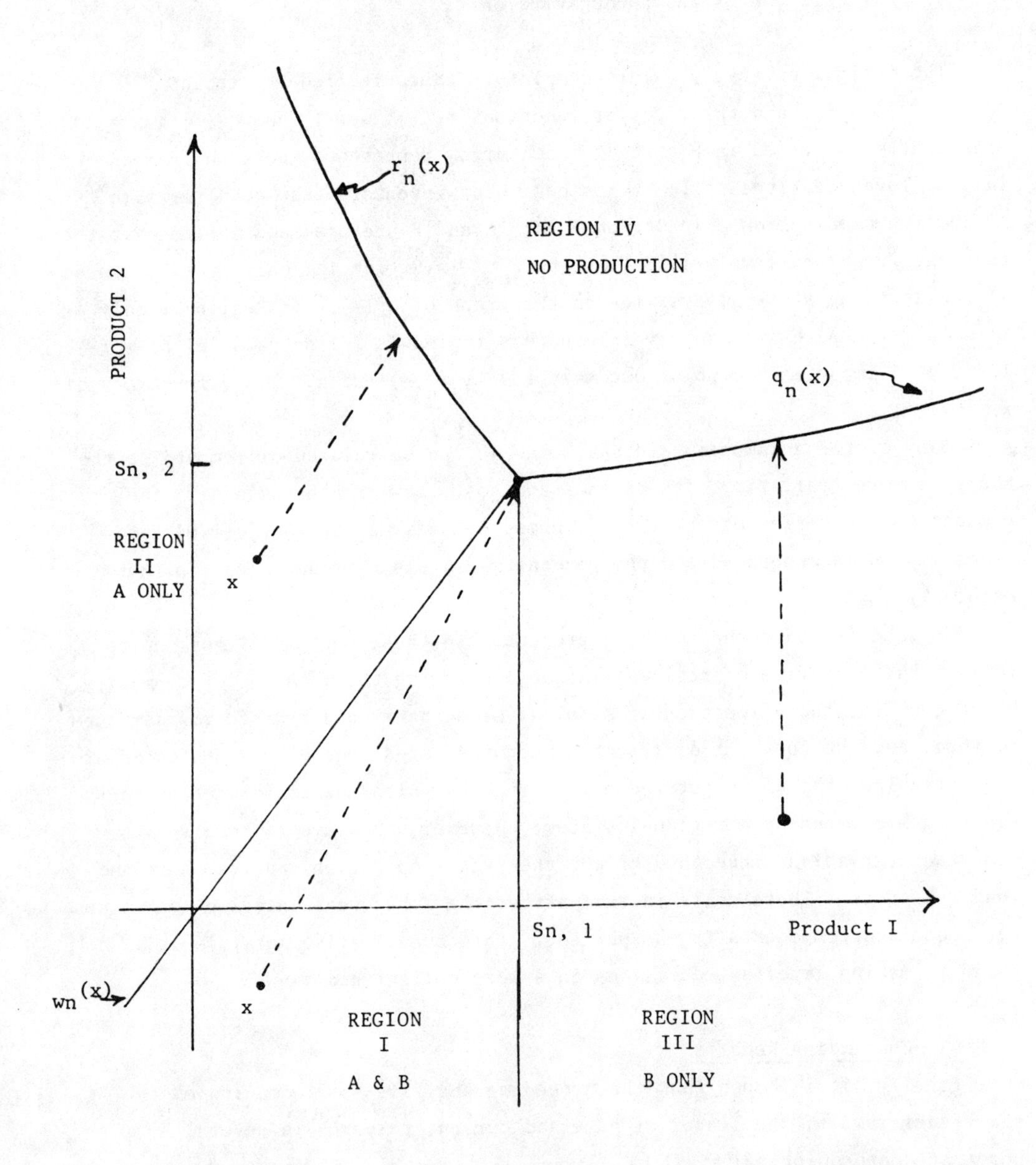

FIGURE 6

Characterization of the Optimal Policy in Period n

REFERENCES

Arrow, K.J., S. Karlin and H. Scarf, eds. Studies in the Mathematical Theory of Inventory and Production, Stanford University Press, Stanford, Calif. (1958).

Barlow, R., L. Hunter and F. Proschan, "Optimum Checking Procedures," J. Soc. Ind. Appl. Math., 11, 4 (1963), pp. 1078-1095.

Calvo, A.B. and D.H. Marks, "Location of Health Care Facilities: An Analytical Approach," Socioeconomic Planning Science, 7 (1973), pp. 407-422.

Cohen, M.A. and W.P. Pierskalla, "Managment Policies for a Regional Blood Bank," Transfusion 15 (1975), pp. 58-67.

Cohen, M.A. and W.P. Pierskalla, H.S. Yen, I. Or, "Regionalization and Regional Management Strategies in the Blood Banking System: An Overview." Proceedings of the Second National Symposium on the Logistics of Blood Transfusion Therapy, June 19-20, 1975. Sponsored by the Michigan Association of Blood Banks and the Michigan Community Blood Center.

Cooper, L., "Location-Allocation Problem," Operations Research, 11 (1963).

Cooper, L., "Hueristic Methods for Location-Allocation Problems," SIAM Review, 6 (1964).

Cooper, L., "The Transportation-Location Problem," Operations Research, 20, (1972).

Derman, C., "On Minimax Surveillance Schedules," Naval Research Logistics Quarterly, 8 (1961), pp. 415-419.

Deuermeyer, B. and W.P. Pierskalla, "A By-Product Production System With An Alternative," Management Science, 24, (1978), pp. 1373-1383.

Eddy, David M., "Screening for Cancer: Theory, Analysis, and Design," Department of Family, Community, and Preventive Medicine, Stanford University, (1978).

Eddy, David M., "Implementation Plan: Part III. Rationale for the Cancer Screening Benefit Program Screening Policies," Department of Family, Community, and Preventive Medicine, Stanford University, for Blue Cross Association, (1978).

Francis, R.L., and J.A. White, Facility Layout and Location, Prentice-Hall, Inc., Englewood Cliffs, N.J. (1974).

Fries, B., "Optimal Ordering Policy for a Perishable Commodity With Fixed Lifetime," Operations Research, 23 (1975), pp. 46-61.

Hutchinson, G. and S. Shapiro, "Lead Time Gained by Diagnostic Screening for Breast Cancer," Journal of the National Cancer Institute, 41 (1968), p. 665.

Jennings, J., "Blood Bank Inventory Control," Management Science, 19 (1973), pp. 637-645. Also appears in Analysis of Public Systems, R. Keeney, P. Morse (editors), MIT Press, Cambridge, Mass., (1972).

Jennings, J., "Inventory Control in Regional Blood Banking Systems," Technical Report Number 53, Operations Research Center, MIT (1970).

Keller, J., "Optimum Checking Schedules for Systems Subject to Random Failure," Management Science, 21 (1974), pp. 256-260.

Kirch, R. and M. Klein, "Surveillance Schedules for Medical Examinations," Management Science, 20 (1974), pp. 1403-1409.

Kirch, R. and M. Klein, "Examination Schedules for Breast Cancer," Cancer, 33 (1974), pp. 1444-1450.

Kolesar, P., "A Markovian Model for Hospital Admission Scheduling," Management Science, 16, 6 (February 1970), pp. B-384-396.

Lincoln, T., and G.H. Weiss, "A Statistical Evaluation of Recurrent Medical Examinations," Operations Research, 12 (1964), pp. 187-205.

McCall, J., "Maintenance Policies for Stochastically Failing Equipment: A Survey," Management Science, 11 (1965), pp. 493-524.

McCall, J., "Preventive Medicine Policies," Rand Corporation, P-3368-1, (1969).

Magnanti, T., B.L. Golden, and H.Q. Nguyen, "Implementing Vehicle Routing Algorithms," Technical Report No. 115, Operations Research Center, Massachusetts Institute of Technology (September 1975).

Miller, H.E., W.P. Pierskalla, and G.J. Rath, "Nurse Scheduling Using Mathematical Programming," Operations Research, 24, 5 (September-October 1976), pp. 857-870.

Nahmias, S. and W.P. Pierskalla, "Optimal Ordering Policies for a Product That Perishes in Two Periods Subject to Stochastic Demand," Naval Research Logistics Quarterly, 20 (1973), pp. 207-229.

Nahmias, S., "Optimal Ordering Policies for Perishable Inventory," Operations Research, 23 (1975), pp. 735-749.

Nahmais, S., "Inventory Depletion Management When the Field Life is Random," Management Science, 20 (1974), pp. 1276-1283.

Nahmais, S. and W.P. Pierskalla, "A Two-Product Perishable/Nonperishable Inventory Problem," SIAM J. Applied Math., 30, No. 3 (1976), pp. 483-500.

Or, I., "Traveling Salesman-Type Combinatorial Problems and Their Relation to the Logistics of Regional Blood Banking." Doctoral Dissertation, Northwestern University, Evanston, Illinois - 60201 (1976).

Or, I., and W.P. Pierskalla, "BTAP: A Computer Program to Obtain Solutions to the Blood Transportation-Allocation Problem and Other Traveling Salesman Type Problems," Technical Report, Department of Industrial Engineering and Management Sciences, Northwestern University, Evanston, Illinois (August 1976).

Or, I., and W.P. Pierskalla, "A Transportation Location-Allocation Model for Regional Blood Banking," AIIE Transactions, (Summer 1979).

Pierskalla, W.P. and J.A. Voelker, "A Survey of Maintenance Models: The Control and Surveillance of Deteriorating Systems," Naval Research Logistics Quarterly, 23, 3 (1976), pp. 353-388.

Pierskalla, W.A. and J.A. Voelker, "A Model for Optimal Mass Screening and the Case of Perfect Test Reliability," Technical Report No. 3, Department of Industrial Engineering and Management Science, Northwestern University, Evanston, Illinois - 60201 (1977).

Pierskalla, W.P. and C. Roach, "Optimal Issuing Policies for Perishable Inventory," Management Science, 18 (1972), pp. 603-615.

Pierskalla, W.P. and J.A. Voelker, "Mass Screening With a Heterogeneous Population," Working Paper, Department of Industrial Engineering and Management Sciences, Northwestern University, Evanston, Illinois (1978). Submitted to Operations Research.

Prastacos, G.P. and E. Brodheim, "A Regional Blood Distribution Model," Working Paper, The Wharton School, University of Pennsylvania, Philadelphia, Pennsylvania (1977).

Prastacos, G.P., "Allocation of Perishable Inventory," Technical Report No. 70, Operations Research Group, Columbia University, New York, New York (January 1977).

Prorok, P., "On the Theory of Periodic Screening for the Early Detection of Disease," Unpublished Ph.D. Dissertation, University of New York at Buffalo (1973).

Prorok, P., "The Theory of Periodic Screening, I: Lead Time and Proportion Detected," Advances in Applied Probability, 8 (March 1976), pp. 127-143.

Prorok, P., "The Theory of Periodic Screening II: Doubly Bounded Recurrence Times and Mean Lead Time and Detection Probability Estimation," Advances in Applied Probability, 8 (September 1976), pp. 460-476.

Roeloffs, R., "Minimax Surveillance Schedules With Partial Information," Naval Research Logistics Quarterly, 10 (1963), pp. 307-322.

Roeloffs, R., "Minimax Surveillance Schedules for Replacement Units, Naval Research Logistics Quarterly, 14 (1967), pp. 461-471.

Schwartz, M. and H. Galliher, "Analysis of Serial Screening in an Asymptomatic Individual to Detect Breast Cancer," Technical Report, Department of Industrial and Operations Engineering, College of Engineering, University of Michigan (1975).

Thompson, D. and R. Disney, "A Mathematical Model of Progressive Diseases and Screening," presented at the November 1976 ORSA/TIMS Meeting.

Vienott, A.F., Jr., "Optimal Policy for a Multi-Product, Dynamic, Non-stationary Inventory Problem," Management Science, 12 (1965), pp. 206-222.

Vienott, A.F., Jr., "The Status of Mathematical Inventory Theory," Management Science, 12 (1966), pp. A745-A777.

Voelker, J.A., "Contributions to the Theory of Mass Screening," Ph.D. Dissertation, Northwestern University, 1976.

Wendell, R.E. and A.P. Hurter, "Location Theory, Dominance, and Convexity," Operations Research, 21 (1973), pp. 314-320.

Wendell, R.E. and A.P. Hurter, "Optimal Locations on a Network," Transportation Science, 7 (1973), pp. 18-33.

Yen, H., "Inventory Management for a Perishable Product Multi-Echelon System," Ph.D. Dissertation, Northwestern University, 1975.

Zelen, M. and M. Feinleib, "On the Theory of Screening for Chronic Diseases," Biometrika, 56 (1969), p. 601.

Zelen, M., "Problems in the Early Detection of Disease and the Finding of Fault," presented at the meeting of the International Statistics Institute, Washington, D.C. (1971).

Proceedings of Symposia in Applied Mathematics
Volume 25, 1981

OPERATIONS RESEARCH: APPLICATIONS IN AGRICULTURE

Robert B. Rovinsky
U.S. Department of Agriculture

Christine Shoemaker
Cornell University

ABSTRACT: Operations Research techniques have been applied to a wide range of problems arising in agriculture. After surveying applications to a variety of agricultural problems, we focus on a review of mathematical models used for pesticide policy analysis.

In the following discussions we will distinguish between two types of agricultural problems, which differ both in the role mathematical researchers can take and in the methodologies they may employ. The first are "farm level" problems, a term which has traditionally been used to describe decisions made at the local, basic unit of agriculture-the farm. However, the definition may be broadened to include decisions by groups, including individual agri-business firms, farm cooperatives, and forest managers. In these instances the "goals" of the decisionmaker are usually clear and can often be mathematically specified, as can the constraints and conditions under which the system works. The form of analysis will depend upon such things as the level of uncertainty, the boundary conditions, and the data available. The second type of agricultural problem concerns "policy." In such problems there may be several competing decisionmakers, each with different goals. In addition, it may not be possible to precisely specify the constraints and other relationships among variables.

Table 1 lists a few of the applications of operations research in agriculture. Day and Sparling (1977), Johnson and Rausser (1977), and Rausser et al. (1980) give a more detailed review. Their articles contain selected bibliographies totalling over 500 published applications of Operations Research in Agricultural Economics.

In what follows, we will focus on mathematical models and Operations Research techniques used in pest management, both at the local level where insect and weed damage must be controlled by a systematic use of chemical

1980 Mathematics Subject Classification 90B99

and/or biological controls, and at the regional or national level, where economic and environmental policy issues must be examined using large-scale models of entire sectors of the agricultural system. We will confine our discussion to three recent studies in which we have been heavily involved. The first concerns applications of optimization methods, especially dynamic programming, to pest management. The second involves the use of quadratic programming to study how cooperation among farmers can increase their profits and reduce their risk under a variety of pest management policies, and the third focusses on the evaluation and selection of policy models by the Department of Agriculture in its attempt to respond to environmental restrictions on pesticide use.

Table 1: Examples of Applications

Farm Level Applications

Type of Problem	Notes on Technique and Applications
Farm Management	
Diet Selection	-First formulated in 1941, solved by Linear Programming (LP) in 1947. for animals in 1951, now a routine service available at Agriculture Experiment Stations.
Crop Rotation	-One of the first applications of linear programming.
Comparative Static Analysis	-Parametric programming
Farm Budgets, Economics of Soil Conservation	-General mathematical programming, including integer, nonlinear, mixed integer.
Transportation	-Many types of analyses including network theory. International trade flows are a current problem area.
Farm Investment	-Dynamic programming, multiple optimal replacement theory.
Risk	-Stochastic programming
Weather	-Game theory, probabilistic modelling

(Also, computer-aided Real Time Management Systems are now available at Purdue and Michigan State Universities, and are used by Extension Service personnel.)

Type of Problem	Notes on Technique and Applications
Farm Firm Analysis	
Farm Growth Models, Effect of Development Policies	-Recursive programming (a variation of linear programming; a series of LP problems are solved until a desired economic equilibrium is achieved.)

Policy Applications

Type of Problem	Notes on Technique and Applications
Production Response	
Regional Adjustment	-Linear programming, cluster analysis, dual linear programming.
Dynamic Allocation	-Recursive linear programming
Short-run national planning	-Multisector recursive linear programming.

Interregional and Spatial Economics

Distribution and Pricing	-Early use of Hitchcock-Koopmans transportation model.
Efficient allocation Optimal Location of Processing Plants National Allocation of Resources	-LP and network theory
Spatial Equilibrium Prices	-Parametric and reactive programming
Market Demand and Prices in Spatial Equilibrium	-Quadratic programming

Agricultural Development

Foreign exchange maximization	-Input/output analysis
Large scale development	-Mixed integer LP, decomposition LP, simulation.
Small scale development	-Game theory, linear programming

Applications of Optimization Methods to Pest Management

Crop losses due to pest damage (primarily insects and weeds) have been estimated by the U.S. Department of Agriculture to be 30 percent of production, or about 55 billion dollars in 1974. This is in spite of the fact that almost one billion pounds of pesticides, worth over one billion dollars, were used in that year. The pest problem is even more serious in tropical countries, where disease and malnutrition are directly related to insect and weed infestations. The environmental effects of pesticides and the health hazards they pose to farmworkers have been well documented. Several major pesticides, including DDT, have been withdrawn from use in the United States for health or environmental reasons.

In an attempt to reduce pesticide use and the associated environmental hazards, agriculturalists have utilized a number of non-chemical pest control methods. Cultural pest control methods like destruction of crop remains and changes in harvest dates have been encouraged. The use of biological control agents like insect predators or parasites have also been an important part of pest control programs. The co-ordination of cultural, biological, chemical and other means of pest control is called integrated pest management. Mathematical analysis has been used as part of interdisciplinary research investigations to determine the best ways to integrate the timing and scope of available pest control alternatives.

The dynamics of plant and pest growth and damage levels are complicated. For example, the optimal timing of an insect pest management strategy strongly depends on the age distribution of the pest population, densities of beneficial and pest population, condition of the crop, and temperature, among other factors. One approach has been to develop mathematical simulation models, which attempt to synthesize crop and pest behavior with mathematical equations relating each component of the ecosystem. These models require estimation of three related components--the growth or yield of the crop, the growth or decline of the destructive pest population, and the control, whether it be pesticide applications or the behavior of a prey or parasite species.

For several years, there has been a considerable interest in the use of optimization methods to aid in the development of efficient pest management programs. Given the complexity of the agricultural ecosystem and the stochastic nature of important driving variables such as weather, the simulation models which have been developed to describe the effect of management on the dynamics of the pest and/or the crop typically have a very large number of variables. Because of the cost of solving a simulation model for each of the large number of combinations of control decisions that are possible in pest management analysis, optimization methods have been used to aid in the selection of control policies. The basic mathematical challenge associated with the application of optimization methods to this problem is the development of computationally feasible procedures for such large, complex systems.

Discussions of the applications of optimization methods to pest management analysis have been included in review articles by Jaquette (1972), Conway (1977), and Wickwire (1977). These applications have utilized a variety of optimization methods including a gradient and quadratization technique, Powell's method, a reduced gradient method (Regev et al., 1977), and dynamic programming (Shoemaker, 1973, 1977; Taylor and Headley, 1975). Dynamic programming (Bellman and Dreyfus, 1962) is the technique which has been most widely used because it is especially well suited for analysis of pest control decisions which are made at discrete points in time in a stochastic, observable environment. However, each of the optimization procedures have different advantages and disadvantages. Hence the selection of an optimization method depends upon the mathematical structure of the pest management model. For the sake of brevity we will restrict our review below to several recent applications of dynamic programming. These applications illustrate the procedures used to circumvent the dimensionality problems associated with the application of optimization methods to pest management.

Dynamic programming is an optimization technique which is based upon the solution of recursive equations describing the decision process at each of a number of stages. For a stochastic decision problem the form of the recursive equation is:

$$H^k(\underset{\sim}{s}^k) = \underset{\underset{\sim}{v}^k}{\text{Min}}[R(\underset{\sim}{s}^k, \underset{\sim}{v}^k) + a \sum_{i=1}^{L} P[\underset{\sim}{s}^{k+1} = \underset{\sim}{\omega}_i | \underset{\sim}{s}^k, \underset{\sim}{v}^k] H^{k+1}(\underset{\sim}{\omega}_i)]] \quad (1)$$

where $\underset{\sim}{s}^k$ is the state vector in period k, $\underset{\sim}{v}^k$ is the decision vector and $H^k(\underset{\sim}{s}^k)$ is the minimum cost of going from period k to the end of the decision horizon. $R(\underset{\sim}{s}^k, \underset{\sim}{v}^k)$ is the net return (e.g., profit minus cost) received in period k. $P[\underset{\sim}{s}^{k+1} = \omega_i | \underset{\sim}{s}^k, \underset{\sim}{v}^k]$ is the probability that the state $\underset{\sim}{s}^{k+1}$ will equal ω_i given that the state in period k is $\underset{\sim}{s}^k$ and the control decision $\underset{\sim}{v}^k$. The constant a is a discount factor. In the deterministic case, the equation can be more simply written as:

$$H^k(\underset{\sim}{s}^k) = \underset{\underset{\sim}{v}^k}{\text{Min}}[R(\underset{\sim}{s}^k, \underset{\sim}{v}^k) + a\, H^{k+1}(\underset{\sim}{G}(\underset{\sim}{s}^k, \underset{\sim}{v}^k))] \quad (2)$$

where the ith component of the state vector $\underset{\sim}{s}_i^{k+1} = G_i(\underset{\sim}{s}^k, \underset{\sim}{v}^k)$. The components of the state vector s^k are usually used to describe the age structure of the plant or pest population, as well as variations in spatial distributions or genetic composition.

One pest which has been the object of several optimization studies is the spruce budworm. Epidemic outbreaks of this pest, which occur approximately every forty or fifty years, have resulted in tree mortality on hundreds of thousands of acres in eastern Canada and the United States. Large simulation models have been created to describe the effect of pest control practices on the occurrence of outbreaks and on tree mortality (Jones, 1977; Stedinger, 1977). These models are very complex, involving several hundred geographical locations, and over twenty variables at each site to describe budworm densities, tree damage and tree age structure. The time steps of the models are one year, and the model is computed for at least one hundred years to evaluate the impact of alternative control strategies.

Attempts to utilize optimization methods for analyzing spruce budworm management have required a reduction in the number of state variables over those used in the simulation models. Using a method suggested by Dantzig (1974), Winkler (1975) developed a dynamic programming model with two state variables. The first variable s_1^k is the density of spruce budworm eggs at the start of the kth year and s_2^k is a measure of the amount of defoliation of the trees which had occurred in the previous year. The problem formulation is similar to that given in Equation (1), where the probability P reflects the impact of stochastic weather events on budworm reproductive and mortality rates. In year 1 all trees are assumed to be one year old. The two most important factors which are ignored in this description are the mixed age structure of the forest and the spatial distribution of budworm densities. As pointed out by Holling and Dantzig (1978), this formulation does not incorporate the fact that a pesticide application cannot be directed only at trees of a fixed age, but rather it must be applied to an entire stand, in which tree ages usually vary over a considerable range.

Because optimization methods require a relatively simple description of a system, it is advisable to test the policies selected by optimization models with simulation models which incorporate a more detailed description of the system. Holling and Dantzig (1978) tested the policy generated by the Dantzig-Winkler algorithm with the simulation model developed by Jones (1977). The results of the spatially disaggregated simulation model indicated that profits, total wood supply and employment over a one hundred year planning horizon are greater with the Dantzig-Winkler policy than with the policy which has historically been used in New Brunswick. Although the Dantzig-Winkler policy appears to perform better than the current policy, it is unlikely to be a true optimum since the optimization formulation could not incorporate a realistic description of the impact of the age composition of trees and spatial heterogeneity of the pest density.

In order to overcome some of the simplifications associated with the Dantzig-Winkler policy, Stedinger (1977) developed an alternate formulation. Initially Stedinger considered a single site of trees of mixed ages under steady-state conditions. He assumed the age distribution of the trees was stable, *i.e.*, that 1/60 of the forest consisted of trees which were i years old, $i = 1, \ldots, 60$. To minimize the infinite-horizon, discounted costs of pesticide spraying, Stedinger solved the following recursive dynamic programming equation:

$$H_{\infty}(\underset{\sim}{s}) = \min_{\underset{\sim}{v}(\underset{\sim}{s})} \left[R(\underset{\sim}{s}, \underset{\sim}{v}) + a \sum_{i=1}^{L} P(\underset{\sim}{\omega}_i | \underset{\sim}{s}, \underset{\sim}{v}) \, H_{\infty}(\underset{\sim}{\omega})\right] \tag{3}$$

where s_1 is the egg mass density and s_2 is an index of deformation. Stedinger calculated the transition probabilities $P(\underset{\sim}{\omega}_1 | \underset{\sim}{s}, \underset{\sim}{v})$ from his own detailed simulation model of spruce budworm population dynamics in Maine. He used Howard's (1960) policy iteration procedure to compute the solution to (3).

In order to insure that the stable age distribution of trees was maintained, Stedinger could not allow tree mortality to occur; hence he constrained v to equal one when egg densities or defoliation are sufficiently high. This is a very restrictive assumption since it may be more economical to let trees die in some areas rather than to spray them year after year. Stedinger also compared the output of single site and multi-site models for a variety of control strategies. He concluded that the single site management model used both by himself and by Winkler (1975) is not adequate because it ignores the economic value of spraying moderate or high density populations in order to prevent infestation of adjacent sites.

In many cases, especially in control of insect pests of field crops, it is very important to incorporate dynamic changes in the age structure of the pest population into pest control decisions. Age distribution is significant because the pest's susceptibility to pesticide, its ability to inflict crop damage and its reproductive potential all depend upon the age of the pest. Inclusion of age class sizes in the state vector of an optimization problem usually increases the dimension of the problem so much that it is not computationally feasible to calculate the optimal solution. This is similar to the problem of describing the age structure of host trees encountered in the studies of spruce budworm.

Shoemaker (1979) suggests a method of circumventing the dimensionality problems associated with the incorporation of pest age structure into a dynamic programming formation. Her approach is to use the state vector to describe the times of previous pesticide applications. The recursive dynamic programming equations are in the form of Equation (2) with the state vector $\underset{\sim}{s} = (s_1, s_2)$ giving the times of the two most recent pesticide applications. The function R is the negative of the cost during period k of crop damage and pesticide applications. The amount of crop damage occurring in a period depends upon the age structure of the population which is computed at each stage from the equation

$$X_i = S(k-i, \underset{\sim}{s})A(k-i)$$

where X_i is the number of pests in the ith age class in period k, A(k-i) is the number of individuals recruited at time i and S $(k-i,\underset{\sim}{s})$ is the fraction of the cohort which survive from period k-i to period k given that the most recent periods of pesticide application are at times $\underset{\sim}{s} = (s_1, s_2)$. Expressing survival S as a function of $\underset{\sim}{s}$ is an approximation since it is possible that more than two pesticide applications have been made. However, because pesticide mortality rates are quite high, the fraction of the population which would survive two pesticide applications is quite small. Therefore, the actual survival can be closely approximated by considering only the two most recent pesticide applications. As a result the state of the system can be characterized by only two variables.

Because of this two dimensional characterization of the pest population, the optimal timing of pesticide applications can be computed more efficiently for population models with a large number of age classes than is possible with the methods used previously. The two dimensional characterization is based on a number of assumptions including restrictions that survival and damage are deterministic functions and that seasonal damage can be represented as a sum of the damage occurring in each stage. It is also possible to incorporate non-linear damage or stochastic effects by inclusion of an additional state vector (Shoemaker, in press). Solution of the associated three or four dimensional dynamic programming problem requires more computation time than in the linear, deterministic case, but even for a large number of age classes these computations are quite feasible.

Another paper by Shoemaker (1977) develops a procedure for solving an optimization model which incorporates biological and cultural pest control procedures as well as the use of pesticides and a dynamic description of pest age structure. The inclusion of biological control provided by an insect parasite considerably complicates the calculation of optimal policies because pesticide and cultural control procedures may kill parasites as well as pest insects. The approach used by Shoemaker is to develop two nested models. The first model is a multi-year, stochastic dynamic programming model with three state variables. The recursive equations of this model have the form of Equation (1), where the decision vector v describes the times of harvest (v_1) and the dosage of pesticide application (v_2). The components of the state vector are the number of pests (s_1) and parasites (s_2) at the beginning of the season and the weather pattern which is described by a random variable s_3. The variable s_3 is important because synchrony between the crop, the pest and parasite populations can be significantly altered by changes in temperature.

The second model is a more detailed population model which is used to estimate the effect of management decision $\underset{\sim}{v}^k$ and state vector $\underset{\sim}{s}^k$ on yield losses and on the size of next year's populations (s_1^{k+1}, s_2^{k+1}). The population model incorporates the description of the effect of temperature on dynamic changes in pest age structure, on crop susceptibility and on synchrony between the occurrence of parasites and susceptible pest age classes. A detailed population model describing these interactions could have been simulated to calculate the decision model's transition functions $G_1(\underset{\sim}{s}^k, \underset{\sim}{v}^k) = s_1^{k+1}$ and $G_2(\underset{\sim}{s}^k, \underset{\sim}{v}^k) = s_2^{k+1}$ and the return function $R(\underset{\sim}{s}, \underset{\sim}{v})$. Unfortunately there are so many combinations of values of $(\underset{\sim}{s}, \underset{\sim}{v})$ that simulation of the population model for each value of $(\underset{\sim}{s}, \underset{\sim}{v})$ is not feasible.

To avoid the computational difficulties associated with trying to compute a simulation model for each value of $\underset{\sim}{s}$ and $\underset{\sim}{v}$, Shoemaker developed a differential equation model which mimicked the most essential features of existing simulation models. The advantage of the differential equation model is that it is possible to solve parts of it analytically.

The basis of the population model is an equation describing the effect of predation and other mortality factors on the number of unparasitized insects ($N_n(r,b)$) born at a particular time (b) and surviving to at least level r of maturation

$$\frac{dN_n}{dr}(r,b) = - D(r-b)N_n(r,b) - \alpha(r-b)N_n(r,b)P(h(r))\dot{h}(r) \quad (4)$$

where D is the death rate, P(t) is the number of parasites attacking at time t and α is the rate of attack. The number of active parasites at any point in time is $P(t) = s_2^k P_o(t)$ where s_2^k is the total number of parasites and $P_o(t)$ describes their distribution through time. The parasite only attacks certain age classes so α has the form

$$\begin{aligned} \alpha(m) &= \alpha_1 \quad \text{if } A_1 \le m \le A_2 \\ &= 0 \quad \text{otherwise} \end{aligned} \quad (5)$$

where m is the maturity of the individual. The pest population matures at a rate that is dependent primarily on temperature. The activity of the parasite is more closely related to the passage of calendar time. The index t is used to denote the maturation time scale of the parasite and r refers to the maturation scale of the pest. The function h(r)=t relates the two time scales. Computation of h(r) depends upon temperature, and it is defined in such a way that the maturity m of $N_n(r,b)$ is r-b. The term $\dot{h}(r)$ is required in Equation (4) to convert the parasitism rate α (which is in units of t) to a rate in terms of units of r.

Following a similar argument, the number of parasitized insects is

$$\frac{dN_p}{dr}(r,b) = -D(r-b)N_p(r,b) + \alpha(r-b)N(r,b)P(h(r))\dot{h}(r) \cdot \quad (6)$$

Since all eggs are unparasitized, $N_p(b,b) = 0$ for all b. The number of new eggs being laid, $N_n(b,b)$, is specified in an input function θ which describes the distribution of oviposition over time, i.e., $N_n^k(b,b) = s_1^k\,\theta(b)/2$.

The solution to Equation (4) is

$$N_n(r,b) = N_n(b,b)\exp[-\int_b^r [D(\tau-b) + \alpha(\tau-b)P(h(\tau))\dot{h}(\tau)]d\tau] \quad (7)$$

$$= s_1^k \frac{\theta(b)}{2} \exp(-K(s-b))\exp(-\alpha_1 \int_{\underline{r}}^{\bar{r}} P(h(s))\dot{h}(s)ds) \quad (8)$$

where

$$K(m) = \int_o^m D(\tau)d\tau , \quad (9)$$

$\underline{r}(b,r) = \mathrm{Min}(b+A_1,r)$ and $\bar{r}(b,r) = \mathrm{Min}(b+A_2,r)$.

The interval $(\underline{r}(b,r), \bar{r}(b,r))$ is the period preceding r when the individual is susceptible to parasitism, i.e, when $\alpha(r-b) = \alpha_1$. The integral in Equation (8) can be replaced by

$$L(b,r,s_2,s_3) = \int_{\underline{t}}^{\bar{t}} P(t)dt \qquad (10)$$

where $\bar{t} = h(\bar{r}(b,r))$ and $\underline{t} = h(\underline{r}(b,r))$. The function $h(r)$ is determined by the random state variable s_3 which depends upon weather.

From the equations above, it follows that in the absence of pesticide applications

$$N_n(r,b) = s_1^k \frac{\theta(b)}{2} R(r-b)\exp\{-\alpha_1 L(b,r,s_2,s_3)\} \qquad r \geq b \qquad (11)$$

where $R(a) = \int_o^a K(\tau)d\tau$.

Harvest and pesticide mortality is modelled by

$$N_n(v_1^+,b) = \phi(v_2,v_1-b)N_n(v_1^-,b) \qquad (12)$$

where v_1 is the time of harvest, v_2 is the insecticide dosage, and $\phi(v,m)$ is the survival from pesticide and harvesting. Then the general expression is

$$N_n(r,b) = N_n(b,b)R(r-b)\hat{\phi}(\underset{\sim}{v},b,r)\exp\{-\alpha_1 L'(b,r,s_2,s_3)\} \qquad (13)$$

where L' is adjusted to include the effects of pesticides on parasites and

$$\hat{\phi}(\underset{\sim}{v},b,r) = \phi(\underset{\sim}{v},b) \qquad \text{if } r \geq v_1$$
$$= 1 \qquad \text{otherwise} \qquad (14)$$

The dynamic programming model requires evaluation of the functions G_1 and G_2 which describe year-to-year changes in the values of the state vector. The number of pests (s_1^{k+1}) at the beginning of year k+1 is the fraction (μ_w) which survive the overwintering period multiplied by the number which have survived to the end of (r_ε) of the previous growing season, which is $\int N_n(r_\varepsilon,b)db$. Combining the relationships described above, we can show that

$$s_1^{k+1} = s_1^k \frac{R(Amax)}{G}\mu_w \int M(b,\underset{\sim}{v})\exp\{-\alpha_1 L'(b,r,s_2^k,s_3^k)\}\, db$$
$$= G_1(s_1^k,s_2^k,s_3^k,v_1^k,v_2^k) \qquad (15)$$

where

$$M(b,\underset{\sim}{v}) = \frac{\theta(b)}{2}\phi(v_2,v_1-b) \qquad (16)$$

and A_{max} is the age at which pests enter their overwintering stages. Similarly the number of parasites in the following year is

$$s_2^{k+1} = s_1^{k+1} \frac{R}{2}(A_p)\mu_p \int M'(b,\underset{\sim}{v})[1-\exp\{-\alpha_1 L'(b,r,s_2^k,s_3^k)\}\, db$$

$$= G_2(\underset{\sim}{s}^k,\underset{\sim}{v}^k) \qquad (17)$$

where

$$M'(b,\underset{\sim}{v}) = M(b,\underset{\sim}{v}) \qquad \text{if } v_1 - b < A_p$$

$$= \gamma\frac{\theta(b)}{2} \qquad \text{otherwise} \qquad (18)$$

The scalar A_p is the age at which parasites kill their hosts and become pupae. The pupae are not susceptible to pesticide but a fraction γ will be killed by harvesting. The scalar μ_p is the fraction of parasites which survive the winter.

Since the function L and the integrals must be solved numerically, Equations (15) and (17) do not represent an analytical solution. However, using Equations (4) and (5), a highly efficient numerical procedure was developed which loops through the values of s_1, s_2, s_3, v_1 and v_2 in such a way that solutions for one set of values can be utilized in computing G_1 and R for other sets of values. A description of this procedure and its application to alfalfa weevil management is given in a paper by Shoemaker (in press).

The Optimal Distribution of Crop Production Under Differing Levels of Farmer Cooperation: An Example Using Cotton

An important policy problem in the economics of pest management is how to compute the set of economic equilibria that results from differing policy assumptions. This is particularly important when trying to predict agricultural production and distribution resulting from a change in pesticide use or policy. A variety of linear, quadratic, and concave programming methods have been developed to determine, for a variety of agricultural policies, the optimal location of crops, livestock, and producers (Duloy and Norton, 1975; Hall et. al., 1975; Heady and Srivastava, 1975; Takayama and Judge, 1971; Von Oppen and Scott, 1976). However, previous studies in this field assumed only one type of competition among farmers; either farmers operate only in pure competition as price-takers, or they are assumed in total cooperation. Further, present algorithms require for their solution that the objective function or decision criterion be a concave or convex function. Finally, most policy models in developed nations have been concerned with farm inputs, crop production, or land and water use, and have concentrated on the feed grain and food sector. They exclude as exogenous cotton production, even though over 50% of insecticide use in the U.S. is devoted to this crop. By using a fixed national demand for cotton lint, the models limit cotton land shifts both nationwide and within individual crop regions as well.

Because of these limitations, these models cannot consider such policy issues as changes in insect pest management strategies, irrigation practices, cotton support and set-aside programs, or competition from synthetic fibers. Since it is impossible to eliminate a single fixed demand without considering some type of price structure to control and determine supply, it is necessary to examine models in which both production and price are at least partially variable. Limited work in this area has been done by Casey and Lacewell (1973) and Evans and Bell (1978).

In what follows we present methods and results that relax the constraints discussed above, and greatly extend the methods and applications. We develop and use a new algorithm for solving a particular class of nonconvex quadratic programming problems, which has proven to be quite efficient and inexpensive, and show applications to a recent insect pest management study (Pimentel et al. 1979).

To begin, consider a model where a large number of producers are aggregated into N subsets that we will call regions. Producers in region j can produce either a commodity C (in the study cited above, this is cotton) or an alternative commodity A_j (in the study, soybeans or grain sorghum) whose price is assumed independent of the actions of all other producers. The price of C in region j depends upon the total national production of C, but the regional price of the alternative commodity A_j is assumed independent of the actions of all other producers. This latter assumption is reasonable in this analysis, because the national acreage of cotton is much less than the total acreage of soybeans and grain sorghum, and a majority of acres in these crops are planted outside the principal cotton producing states.

The equilibrium allocation of resources depends upon the amount of cooperation among individual producers and among regions. Let T^L be the total production of commodity C when each individual producer acts independently as a price-taker to maximize his profit. Let T^R be the level of production of commodity C resulting when producers within a region cooperate to maximize profit. Finally, let T^G be the level of production of commodity C when all producers in all regions cooperate to maximize total global (national) profits. The vectors X^L, X^R and X^G give the amounts of commodity C produced in each region for each of these situations.

Mathematical Results

We have elsewhere (Rovinsky, 1977; Rovinsky et al., in press) presented results which yield methods for calculating X^L, X^R, and X^G. These results will be briefly stated and reviewed below.

First, we define the following for each region j:

P_j - price of commodity C

x_j - the number of units of C produced

L_j - total capacity

s_j - profit per unit of capacity for producing alternative commodity A_j

y_j - amount of C produced per unit of capacity

b_j - the maximum production of commodity C ($b_j = L_j y_j$)

c_j - production cost of C per unit of capacity.

Let T denote the total production $\Sigma_j x_j$ of commodity C and assume that T determines the price in each region by the simple linear relationship

$$P_j = d_j T + f_j$$

where $d_j < 0$ and f_j are constants. Then the profit per unit capacity in region j obtained by producing commodity C is $(d_j T + f_j)y_j - c_j$. The profits per unit of capacity are equal for the two commodities when $T = q_j$ where

$$q_j = -m_j/d_j \text{ and } m_j = f_j - (c_j + s_j)/y_j \tag{2}$$

Producing commodity C in region j is more profitable than producing commodity A_j if and only if $T < q_j$.

Result 1: To find X^L, order the regions so that $q_1 \geq \ldots \geq q_N$. Let k_p be the smallest k so that $\sum_{j=1}^{k} b_j > q_k$. If there is no such k, set $k_p = N+1$. Then:

$$X_j^L = \begin{cases} b_j & 1 \leq j < k_p \\ \max\left(0, q_{k_p} - \sum_{j=1}^{k_p-1} b_j\right) & j = k_p \\ 0 & k_p < j \leq N \end{cases} \tag{3}$$

This algorithm is due to Casey and Lacewell (1973). We have shown that, if each producer is acting independently to maximize his profit, then the distribution of production X^L given above is an economic equilibrium; i.e., no producer can independently act to increase his profits. Further, it is the unique such equilibrium.

Result 2: Under regional equilibrium, the total profit in region j is:

$$(d_j \sum_{i=1}^{N} x_i)x_j + m_j x_j + s_j L_j \tag{4}$$

To find X^R, define

$$f(T) = \Sigma\{b_i: 0 \le T \le q_i - b_i\} + \Sigma\{q_i - T: q_i - b_i \le T \le q_i\}. \tag{5}$$

Then it is easy to see that f has a unique fixed point T^R, and X^R is given by

$$x_j^R = \begin{cases} b_j & \text{if } 0 \le T^R < q_j - b_j \\ q_j - T^R & \text{if } q_j - b_j \le T^R \le q_j \\ 0 & \text{otherwise} \end{cases} \tag{6}$$

We have also shown that X^R is the unique regional equilibrium, that the resulting total production $\sum_{j=1}^{N} x_j^R = T^R$, and that $T^R \le T^L$.

Result 3: If we assume cooperation among regions as well as among producers within a region, then the total profits can be further increased. From (4), we note that the total income from commodities C and $A_., A_2, \ldots, A_N$ is

$$\sum_{i=1}^{N} \sum_{j=1}^{N} d_i x_i x_j + \sum_{i=1}^{N} m_i x_i + \sum_{i=1}^{N} s_i L_i, \tag{7}$$

and thus the problem of finding a global equilibrium can be stated as the following nonconcave quadratic programming problem:

$$\max \theta(x) = \sum_{i=1}^{N} \sum_{j=1}^{N} d_i x_i x_j + \sum_{i=1}^{N} m_i x_i \tag{8}$$

subject to $0 \le x_j \le b_j$, $j=1,\ldots,N$.

We have shown that if $T = \Sigma x_j$, $r_j = d_j T + m_j$, and the regions ordered for each T so that $r_1 \ge \ldots \ge r_N$, then we can easily define an optimal $x^*(T)$, by a method similar to the Casey-Lacewell ordering algorithm above. Thus we redefine our problem above to the one-dimensional

$$\max \Theta(T) = \theta(x^*(T)) \tag{9}$$

$$0 \le T \le \sum_{j=1}^{N} b_j$$

We have shown that $\hat{\theta}$ is continuous on $S = [0, \sum_{j=1}^{N} b_j]$, piecewise strictly concave, and twice differentiable on all but a finite, known subset of S. Thus we are led to a fairly straightforward search for T^G in S. The search is significantly faster than the existing nonlinear programming algorithms. Details of the algorithm and a listing of the FORTRAN IV program written to implement it are given in Rovinsky (1977). This algorithm, denoted THETA, also finds the local optimal and regional equilibrium solutions described above.

We can significantly reduce the size of the interval of search by finding upper and lower bounds for T^G. We have shown that if T* is any local maximum of Θ, then $T^* \le T^R$. In particular, $T^G \le T^R$ for any T^G which maximizes $\theta(T)$ for $T \in S$. A lower bound T_{min} has also been found for T^G, and THETA calculates for T^G by finding and comparing all local maxima T_I^* between $\max(0, T_{min})$ and T^R. The optimal distribution X^G is then set equal to $x^*(T^G)$.

As discussed above, the THETA algorithm was developed to allow changes in prices, demands, yields, and costs of cotton production to be incorporated into large scale policy models. Thus, instead of a set of fixed demands, price and demand could vary, and regional production could shift accordingly. The THETA algorithm solves for the equilibrium solutions very inexpensively relative to the size of the problems. For example, these methods were applied as part of a recent study of insect pest management methods. Cost, yield, and price data for cotton and its alternate crops were gathered for a partition of the Southern United States into 300 regions. Cotton production and total farm income were computed for the three levels of producer cooperation discussed above, and for each of fifteen price elasticities used to simulate the short, medium, and long run equilibrium estimates for cotton demand. Less than one minute of IBM 370/168 CPU time was required to calculate the complete set of solutions. This algorithm has been extended to a wider class of nonconcave quadratic programming problems. The critical assumption that revenue per unit of capacity is constant for the alternate commodities may be viewed as assuming that the value of unused capacity is fixed. This constraint can be partially relaxed by placing restrictions on shifts of regional production and/or return to C and the alternate commodities. This allows consideration of those situations in which an existing production distribution may be modified under constrained guidelines.

Applications to National Pesticide Policy Analysis

Starting in the late 1960's, researchers began studying the impact of various pesticide policies upon farm income and production. Early work focussed on possible Environmental Protection Agency (EPA) restrictions on or cancellations of widely used insecticides such as DDT, dieldrin and aldrin. Currently the Department of Agriculture (USDA) responds to an EPA notice that a certain pesticide's registration (or re-registration) is being reviewed, by supplying data on the benefits and costs of the pesticide's use by farmers. At first USDA used a partial budgeting approach. This involved first polling state entomologists and economists to obtain the relative inputs of the affected pesticide and its alternatives in their areas. Then the cost and yield changes obtained were aggregated to derive regional and national effects. This approach drew criticism for its failure to reflect expected adjustments by producers and the market (shifts in crop acreage and prices) in response to the proposed pesticide cancellation.

In the early 1970's, several researchers outside USDA began using existing linear programming (LP) models to examine the effects of pesticide policies. Such models assumed the farm acreage in the country could be divided into several hundred homogenous regions. The cost of production and yield for a given crop were assumed to be constant throughout each region. The decision variables described the number of acres in each region which were devoted to each crop. Additional variables described the amounts of each crop transported to other regions to meet market demand. In order to investigate the spatial effects of a pesticide ban, yield and cost coefficients in the linear programming model were changed to evaluate the cost and effectiveness of pest control strategies which would be substituted for application of the banned pesticides. Then the base LP solution (i.e. that solution obtained by solving the model under present conditions) could be compared against a proposed pesticide ban solution to examine regional shifts in production and acreage. Various accounting schemes can then be used to examine environmental and economic effects.

The primary source of LP models and data was a series of studies beginning in 1954 at Iowa State University under the direction of Earl O. Heady (Heady and Srivastava, 1975). Heady's work established these models as legitimate policy tools, but they proved to be large, not well documented, and expensive, and several smaller variations were developed. All were cost-minimization LP models, with variable costs and fixed demands for food, feed, and fiber. Output included transportation and production costs, acreages, and production by crop and region. Their solutions were interpreted as projections of cropping patterns to the year 1980, 1985, or 2000. They generally showed large shifts of production, emphasizing the decreasing and westward moving trends in agricultural land use.

Several problems arose in using these LP models, and researchers and policy analysts were generally unwilling to accept the massive interregional shifts of production predicted. Modifications of the algorithms were made to constrain acreage shifts using "policy" upper and lower bounds. Penalties were attached to gross shifts from current levels to simulate farmers' reluctance to change, and regions were further disaggregated by soil quality. Current work is directed towards meeting another problem with LP, the absence of demand functions. The use of quadratic programming and the implanting of stepped demand functions have aided the process, as we have discussed in the previous section.

In 1977 the Natural Resources Economics Division (NRED) within USDA's Economic Research Service began to collect and analyze the current state of pesticide models and data. Their goal, in using models, was to increase the reliability of their short term research results on the economic consequences of continuing or discontinuing the use of particular pesticides for particular purposes. They hoped also to decrease the number of hours required to estimate the benefits of each use of a pesticide. Finally, they expected to test the suitability of each model for future pesticide policy analyses.

This effort attracted the interest of many modelers outside USDA, and generated many proposals to adapt a variety of LP, simulation, and econometric models to analyze pesticide policy. The economists in NRED decided to pursue a 3-stage plan: first to use current, inexpensive working models or approaches; then to adapt current, large models; and finally to investigate possible linkages and improved models. Each model was run using a common data base. This consisted of a set of cost and yield changes, by state, resulting from a proposed EPA cancellation of a leading cotton insecticide.

There were 3 initial models: (1) a static partial budgeting accounting system; (2) PESTDOWN, a simplified linear programming model with 8 crops, 135 producing regions and 4 soil classes/region; and (3) POLYSIM, an econometric model built around the USDA 5 year baseline economic forecasts. These approaches met the criteria of data and model accessibility, low cost, and ease of use. The second stage used four models: the current Iowa State University linear programming model; a stochastic version of POLYSIM; a commercial Macro-Econometric model; and a cotton model constructed at USDA. In what follows we briefly describe the structure of each of the models. For details and a full description of how each model was used see Rovinsky et al. (1979).

Both the ISU and PESTDOWN models are total variable cost minimization Linear Programming (LP) models of the following general form:

$$
\begin{aligned}
\text{(P)} \quad \text{minimize} \quad & F = cx \\
\text{so that} \quad & A_1X \geq D \\
& A_2X \leq R
\end{aligned}
$$

where the vector c contains the variable cost of production for each cropping activity, the costs of transporting farm commodities between demand regions, and other model specific costs; the vectors D and R contain the exogeneously determined demands and resources, respectively, and the matrices A_1 and A_2 contain the technical coefficients of production and transportation. The models are both quite large and very expensive to solve initially; thus, subsequent policy runs are made using the base solution as a starting point. This reduces the cost considerably.

Both models contain the same set of seven crops: corn, sorghum, wheat, oats, barley, soybeans, and cotton. The ISU model also includes activities for hay and silage, while PESTDOWN includes rye and peas in its allowable crop mix. Using the FEDS budget generator at Oklahoma State to supply cost and yield data, production activities are defined on 135 producing regions in PESTDOWN and 105 producing areas in the ISU model. Demands for agricultural commodities are described for a set of "consuming" regions, which in PESTDOWN are aggregates of States and in the ISU model are general market areas around principal trade cities.

Exogenously determined changes in production costs and crop yields, resulting from a given pesticide policy, are used to change the base cost and crop yield values utilized in the models. Each change in cost must be expressed in base-year dollars and enters the model as an amount to be added to (or subtracted from) the per acre variable costs of producing a given crop in a given region. Each change in yield must be expressed as a ratio of new regional costs and yields; the model will determine the optimal (cost-minimizing) solution for the new scenario. The solution's output describes total and regional acreage and production by commodity. The impacts of a pesticide ban may then be estimated through a comparison of the base and cancellation runs.

The econometric cotton model is one of several commodity-specific models developed within USDA. It was designed for use in evaluating the effects of policy changes on the U.S. cotton market. The model consists of three basic sections; a price block; a supply block; and a demand block. Each block consists of a set of equations and mathematical identities. There is a total of 30 equations and five identities which exactly determine the 35 endogenous variables, and the model contains approximately 70 exogenous variables. The supply block is separated into four cotton production regions; price variables are common to the supply and demand segments of the model.

The model simulates annual behavior in the cotton market. The coefficients of the model were estimated using ordinary least squares regression over the period 1951 to 1976. The model consists of a system of non-linear equations and is solved using the Gauss-Seidel gradient method. Levels of demand and supply and endogenous cotton price variables are simultaneously determined for each given year. Within blocks of the model, however, equations are solved in a recursive fashion, ending with the determination of a cotton market equilibrium. Exogenously determined values for the regional changes in production costs and yields are used to change key variables in the model for a pesticide policy impact run. These changes enter the model through the regional yield and planted acreage equations.

The cotton model provides five year forecasts of a number of important cotton sector variables. If run twice, once as a base and once after incorporating policy induced changes in efficiency and costs of cotton production, that policy's impact on cotton yield per acre, regional and total planted acres, regional and total production, farm and wholesale price, foreign exchange earnings from cotton exports, total consumption, and consumers' and producers' surpluses may be estimated.

POLYSIM was constructed differently from most economic or econometric simulation models to ensure compatibility with other policy analyses of USDA. The model makes full use of independent forecast data as a reference baseline. Included are the five-year baseline projections of commodity supplies, prices, and utilization made by USDA commodity specialists using formal forecasting models tempered with their own experienced judgments. The projections contain explicit assumptions concerning the rates of change in population, per capita income, consumer preferences, export demand, technology (including crop yields and livestock gains), and other supply and demand shifters. These projections also assume a specific set of agricultural programs.

POLYSIM is a recursive economic simulation model. Ceteris paribus, supply is a function of the previous year's price; the current year price is determined by the level of supply relative to expected demand; and actual quantity demanded is a function of price. The driving mechanisms in the POLYSIM model are the initial and subsequent changes in commodity prices resulting from policy changes. The magnitude of impact is determined by each commodity's own price elasticities, as well as cross supply and demand elasticities.

Commodities included in the model are feed grains, wheat, soybeans, cotton, cattle and calves, hogs, sheep and lambs, chicken, turkeys, eggs, and milk. As indicated earlier, the model is designed to simulate around a set of baseline estimates; these must be available for all years analyzed. To date, most applications have been for a time horizon of three to five years.

Government economists also utilize models developed by the private sector to conduct their research. These include the commonly used products that are commercially marketed by three firms--Chase Econometrics Associates, Inc. (Chase), Data Resources, Inc. (DRI), and Wharton Econometric Forecasting

Associates, Inc. (Wharton). Typical products include all of the following: access to one or more models, data bases and econometric software; receipt of economic and financial reports and forecasts; and some consulting. Although the commercial models are different from each other in many ways, there are enough similarities so that the following discussion of the Chase system will give a general idea of what is available elsewhere.

The two components of the Chase Econometric System most often used for pesticide analyses are the Chase Agricultural Model and the Chase Macroeconomic Model. Changes in base yields and costs of production due to a pesticide regulation are used for input data to the Agricultural Model. These produce output on changes in farm level prices, crop acreage, production, etc. Changes in farm level prices obtained from the Chase Agricultural Model, or other models (such as the CED Cotton Model or POLYSIM which are discussed above) are then used to shock the Chase Macroeconomic Model; these macroeconomic simulations indicate the influence of agricultural events on the general economy. Macroeconomic performance indicators include WPI (the Wholesale Price Index), CPI (the Consumer Price Index), GNP (Gross National Product), etc.

The USDA researchers found, as expected, that the econometric simulation models tended to be less expensive, and more acceptable to other economists, who tended to be suspicious of the LP model's lack of prices and copious detail. They also discovered that, with the possible exception of their Cotton model, there was no adequate regionalization in the econometric models, to reflect yield variations and shifts over time. Perhaps most importantly, their work has stimulated a complete review of their cost and yield data, especially that required for the LP models, for which inaccuracies could cause the model to put all production in one place. Through extensive use of the commodity models, they have encouraged the development of the more integrated, cross-commodity models. Finally, their work was used in a recent draft study on the feasibility of eradicating the Cotton Boll Weevil.

Current work at USDA is aimed at two objectives. There is interest in attacking a wider variety of pest policy problems other than a single pesticide ban. The range of models available, and the fact that they are constantly being improved by their developers, may make it possible to study a range of integrated pest management techniques, and to combine a controlled set of pesticide applications, cropping techniques, early scouting for pest problems, and careful plant breeding to improve pest control and decrease pesticide use. There is also an attempt underway to link the optimization and simulation models, hopefully emerging with a better forecasting and analysis tool. "Rules of thumb," which apply the percentage effects of the normative (LP) models to the base output of the positive (econometric) models, have been proposed, as well as the use of updated sets and stepped demand models.

REFERENCES

1. Agrawal, R.C. and Heady, E.O., Operations Research Methods for Agricultural Decisions, Iowa State University Press, 1972.

2. Bahrami, K., and Kim, M., "Optimal Control of Multiplicative Control Systems Arising From Cancer Therapy," IEEE Trans. Automatic Control, pp. 537-542, 1975.

3. Bellman, R.E. and Dreyfus, S.E., Applied Dynamic Programming, NJ, Princeton University Press, 1962.

4. Bellman, R.E. Dynamic Programming, NJ, Princeton University Press, 1957.

5. Bellman, R.E. and Kalaba, R., "Some Mathematical Aspects of Optimal Predation in Ecology and Boviculture," Proceedings of the National Academy of Sciences, vol. 46, pp. 718-720, 1960.

6. Casey, J.E., Jr., and Lacewell, R.D., "Estimated Impact of Withdrawing Specified Pesticides from Cotton Production," Southern Journal of Agricultural Economics, vol. 5, No. 1, 1973.

7. Casey, J.E., Jr., and Lacewell, R.D., "A General Model for Estimating the Economic and Production Effects of Specified Pesticide Withdrawals: A Cotton Example," Texas A&M University, Environmental Quality Program, Note 15, 1974.

8. Comins, H.N., "The Management of Pesticide Resistance," Journal of Theoretical Biology, vol. 65, pp. 399-420, 1977.

9. Conway, G.R., "Mathematical Models in Applied Ecology," Nature, 269(22), 1977.

10. Dantzig, G.G., "Determining Optimal Policies for Ecosystems," Technical Report, No. 74-11, Stanford University, California, 1974.

11. Day, R.H. and Sparling, E., "Optimization Models in Agricultural and Resource Economics," A Survey of Agricultural Economics Literature, vol. 2, edited by G.C. Judge et al., University of Minnesota, 1977.

12. Duloy, J.H., and Norton, R.D., "Prices and Incomes in Linear Programming Models," Revised Version of Discussion Paper No. 3, Development Research Center, International Bank for Reconstruction and Development (World Bank), New York, 1975.

13. Evans, Sam and Bell, Thomas M., "How Cotton Acreage, Yield, and Production Respond to Price Changes," Agricultural Economics Research, vol. 30, No. 2, April 1978.

14. Fick, G.W., ALSIM I (Level 1) User's Manual, Cornell University, Department of Agronomy, Mimeo 75-20, 1975.

15. Greenberger, M., Crenson, M.A., and Cressey, B.L., Models in the Policy Process. Russell Sage Foundation, New York, 1976.

16. Hall, H.H., Heady, E.O., Stoecher, A., and Sposito, V.A., "Spatial Equilibrium in U.S. Agriculture: A Quadratic programming Analysis," SIAM Review, 17, 2, 323-338, 1975.

17. Heady, E.O., and Srivastava, U.K., Spatial Sector Programming Models in Agriculture. Iowa State University Press, Ames, 1975.

18. Holling, C.S. and Dantzig, G.B., "Determining Optimal Policies for Ecosystems: An Overview of a Case Study," Technical Report, Institute of Resource Ecology, University of British Columbia, Vancouver, B.C., 1978.

19. Holling, C.S., Jones, D.D., and Clark, W.C., "Ecological Policy Design: A Case Study of Forest and Pest Management," Pest Management, G.A. Norton and C.S. Holling (eds.) London, Pergamon Press, 1978.

20. Howard, R., Dynamic Programming and Markov Processes, NY, Wiley, 1960.

21. Hueth, D., and Regev, U., "Optimal Agricultural Pest Management with Increasing Pest Resistance," American Journal of Agricultural Economics, vol. 56, pp. 524-552, 1974.

22. Huffuker, C.B., New Technology of Pest Control, NY, Wiley (in press).

23. Jaquette, D.L., "Mathematical Models for Controlling Growing Biological Populations: A Survey," Operations Research, vol. 20, pp. 1142-1151, 1972.

24. Jones, D.D., "The Application of Catastrophe Theory to Ecological Systems," Simulation in Systems Ecology, G.S. Innis (ed), Simulation Council Proceedings, 1977.

25. LaDue, E.L., Shoemaker, C.A., Russell, N.P., Rovinsky, R.B., and Pimental, D., "The Potential Impact of Cotton Insect Control Technology," Cornell Agricultural Economics Staff Paper No. 79-31, Ithaca, 1979.

26. Petit, M., "The Role of Models in the Decision Process in Agriculture," Decision Making and Agriculture, edited by T. Dams and K.E. Hunt, University of Nebraska Press, 1977.

27. Pimentel, D., Shoemaker, C.A., Ladue, E.L., Rovinsky, R.B., and Russell, N.P., "Alternatives for Reducing Insecticides on Cotton and Corn: Economic and Environmental Impact," Report to the Environmental Protection Agency, 1979.

28. Rausser, Gordon C., Just, Richard E., and Zilberman, Davis, "Prospects and Limitations of Operations Research Applications in Agriculture and Agricultural Policy," California Agricultural Experiment Station Giannini Foundation of Agricultural Economics, Berkeley, March 1980.

29. Regev, U., Gutierrez, A.P., and Feder, G., "Pests as a Common Property Resource: A Case Study of Alfalfa Weevil Control," American Journal of Agricultural Economics, vol. 58, pp. 188-196, 1976.

30. Rovinsky, R.B., Shoemaker, C.A., and Todd, M.J., "Determining Optimal Use of Resources Among Regional Producers Under Differing Levels of Cooperation," Operations Research, Vol. 28, No. 4, July-August 1980.

31. Rovinsky, R.B., Reichelderfer, K.H., Weisz, R.N., and Quinby, W.A., "The Use of Policy Models to Evaluate the Effects of a Pesticide Ban," 1979 (to appear).

32. Shoemaker, C.A., "Optimal Integrated Control of Pest Populations With Age Structure," Operations Research (in press).

33. Shoemaker, C.A., "Optimal Timing of Multiple Applications of Pesticides With Residual Toxicity," Biometrics, December 1979.

34. Shoemaker, C.A., "Optimization of Agricultural Pest Management III: Results and Extensions of a Model," Mathematical Biosciences, vol. 18, pp. 1-22, 1973.

35. Shoemaker, C.A., "Deterministic and Stochastic Analyses of the Optimal Timing of Multiple Applications of Pesticides with Residual Toxicity" Proceedings of Pest Management Modelling Conference, John Wiley and Sons, New York (in press).

36. Shoemaker, C.A., "Pest Management Models of Crop Ecosystems," Ecosystem Modelling in Theory and Practice, edited by Charles A.S. Hall and John W. Day, Jr., John Wiley & Sons, N.Y., 1977.

37. Stedinger, J.R., "Spruce Budworm Management Models," Ph.D. Thesis, Department of Applied Physics and Engineering, Harvard University, Cambridge, 1977.

38. Takayama, T., and Judge, G.G., Spatial and Temporal Price and Allocation Models, North-Holland Publishing Co., Amsterdam, 1971.

39. Taylor, C.R., and Headly, J.C., "Insecticide Resistance and the Evaluation of Control Strategies for an Insect Population," Canadian Entomologist, vol. 107, pp. 237-242, 1975.

40. U.S. Department of Agriculture, "Agricultural Statistics," U.S. Government Printing Office, Washington, D.C. 1975.

41. Von Oppen, M., Scott, J., "A Spatial Equilibrium Model for Plant Location and Interregional Trade," American Journal of Agricultural Economics, 58, 3, 437-445, 1976.

42. Watt, K.E.F., "The Use of Mathematics and Computers to Determine Optimal Strategy and Tactics for a Given Insect Control Problem," Canadian Entomologist, 96, 1964.

43. Waugh, F.V., "Demand and Price Analysis - Some Examples from Agriculture," Technical Bulletin No. 1316, Economic Research Service, U.S. Department of Agriculture, Washington, D.C. 1964.

44. Wickwire, K., "Mathematical Models for the Control of Pests and Infectious Diseases: A Survey," Theoretical Population Biology, vol. 11, pp. 182-238, 1977.

45. Winkler, C., "An Optimization Technique for the Budworm Forest-Pest Model," International Institute for Applied Systems Analysis, RM-75-11, Laxenburg, Austria, 1975.

Proceedings of Symposia in Applied Mathematics
Volume 25, 1981

MATHEMATICAL MODELING APPLIED TO THE RELOCATION OF FIRE COMPANIES

Warren E. Walker
The Rand Corporation

ABSTRACT. In recent years fire departments in urban areas have experienced a sharp increase in demands for their services while their budgets have generally grown at a rate less than that of inflation. Many of these departments have turned to systems analysis for help, realizing that, if they do not use more effectively what resources they have, their level of service will diminish. The Rand Corporation has provided assistance to the New York City Fire Department and others over the past decade, concentrating on deployment policies, which tie available resources to their distribution and movement in the field. This lecture will first provide a brief overview of the deployment policies that have been analyzed with the help of mathematical models. This will be followed by a more complete discussion of one of these policy questions: how should available fire companies be temporarily relocated to provide coverage when many other companies are busy fighting large fires?

I. INTRODUCTION

Fire is the leading cause of catastrophic accidents (those in which five or more people die) in the United States, annually destroys over $4 billion in property, and costs the total economy an estimated $14 billion per year. Moreover, some of the underlying problems are becoming more severe. For example, technological change has created new fire risks, deterioration of inner-city neighborhoods has spawned rising numbers of fires, and the incidence of arson fires has risen dramatically.

At the same time, fire departments in many cities have experienced budget reductions or growth at a rate less than the rate of inflation. If they are unable to use what resources they have more effectively than in the past, their level of service must diminish.

Improved effectiveness, however, is not easy to achieve in fire departments. Fire service management and operations lean heavily on tradition and on rules of thumb, many of which have not changed very much since the turn of

1980 Mathematics Subject Classification 90C50.

the century. This is especially true of fire department deployment policies, which tie available resources to the actual distribution and use of firefighting services in the field. For example, the present number and arrangement of fire companies in most cities are based more on historical factors, such as where volunteer companies were first organized, than on a careful analysis of actual needs. In 1968, a seven-year, multimillion dollar effort was begun that was primarily devoted to developing and testing new models and methods for fire department deployment analysis. The work was performed at The New York City-Rand Institute. The results have been used in over fifty cities throughout the United States [2].*

The general subject of deployment analysis includes a variety of topics, some of which can be addressed using mathematical models and others not. Most of the issues that can be analyzed using mathematical models concern the manner in which fire companies are located and dispatched:

- How many fire companies should be on duty? This may be a planning decision related to the department's budget, or it may concern the appropriate variation in company levels by time of day or by season of the year.
- How many fire companies should be allocated to each region of the city? A simple nonlinear programming model called the Parametric Allocation Model was developed to address this question [4].
- Where should the city's fire companies be located? This question refers to choosing sites for fire stations. A descriptive deterministic model that calculates expected travel times to fires for different arrangements of fire companies was developed. The model is called the Firehouse Site Evaluation Model [10].
- How many fire companies should be sent to an incoming alarm? The answer may depend on the availability of companies at the time of the alarm and what is known about the nature of the incident at the time the dispatch is made. A semi-Markovian decision model was developed that considers explicitly information on the current alarm and expected future events [9]. The model assumes that alarms occur and are extinguished according to random processes. Its state space is the number of companies busy and the potential seriousness of the incoming alarm. The decision variable is the number of companies to dispatch, and the objective function measures travel time to serious fires.
- Which particular fire companies should be dispatched? Most fire departments dispatch the units that are closest to the location of an

*Numbers in square brackets identify references at the end of this paper.

incident. However, using insights from a queueing model, it was found that under certain conditions it pays *not* to dispatch the closest unit [1].

- Which fire companies should be temporarily relocated when a large fire depletes one part of the city of its fire protection? When one large fire or several small fires are being fought in a single region of a city, protection against future fires in the same region is considerably reduced. It is standard practice in most urban fire departments to protect the exposed region by temporarily relocating fire companies from outside the region into some of the vacant firehouses within the region [6].

All of the above issues are discussed in detail in [3]. The remainder of this talk will be devoted to a description of how we dealt with the last issue.

II. THE RELOCATION PROBLEM

Figure 1 presents an illustration of a situation that is, unfortunately, not unusual in New York City and other large cities. Two serious fires break out at about the same time in the South Bronx. Seven ladder companies are involved for several hours fighting the two blazes. This results in a large region being left without a ladder company close by to respond if another fire should break out in the area.

When this happens in a large city the fire department usually temporarily relocates (moves) some fire companies from their firehouses in parts of the city that are still adequately protected to some of the empty houses. In smaller cities it may be necessary to achieve the same effect by borrowing companies temporarily from neighboring communities via a mutual assistance agreement. The purpose of such temporary relocations is clear--to spread out the still available firefighting resources in order to reduce and balance the risks and consequences that would result if other fires occur.

Any relocation method must provide answers to the following four questions:

1. In what situations are relocations to be made?
2. Which of the empty firehouses should be filled?
3. Which of the available fire companies should be moved?
4. To which empty houses should each be moved?

When alarm rates are low, the fires requiring relocations of firefighting units are rare. They occur one at a time, and when they occur, no other fires are typically in progress. Thus, under low alarm rate conditions it is

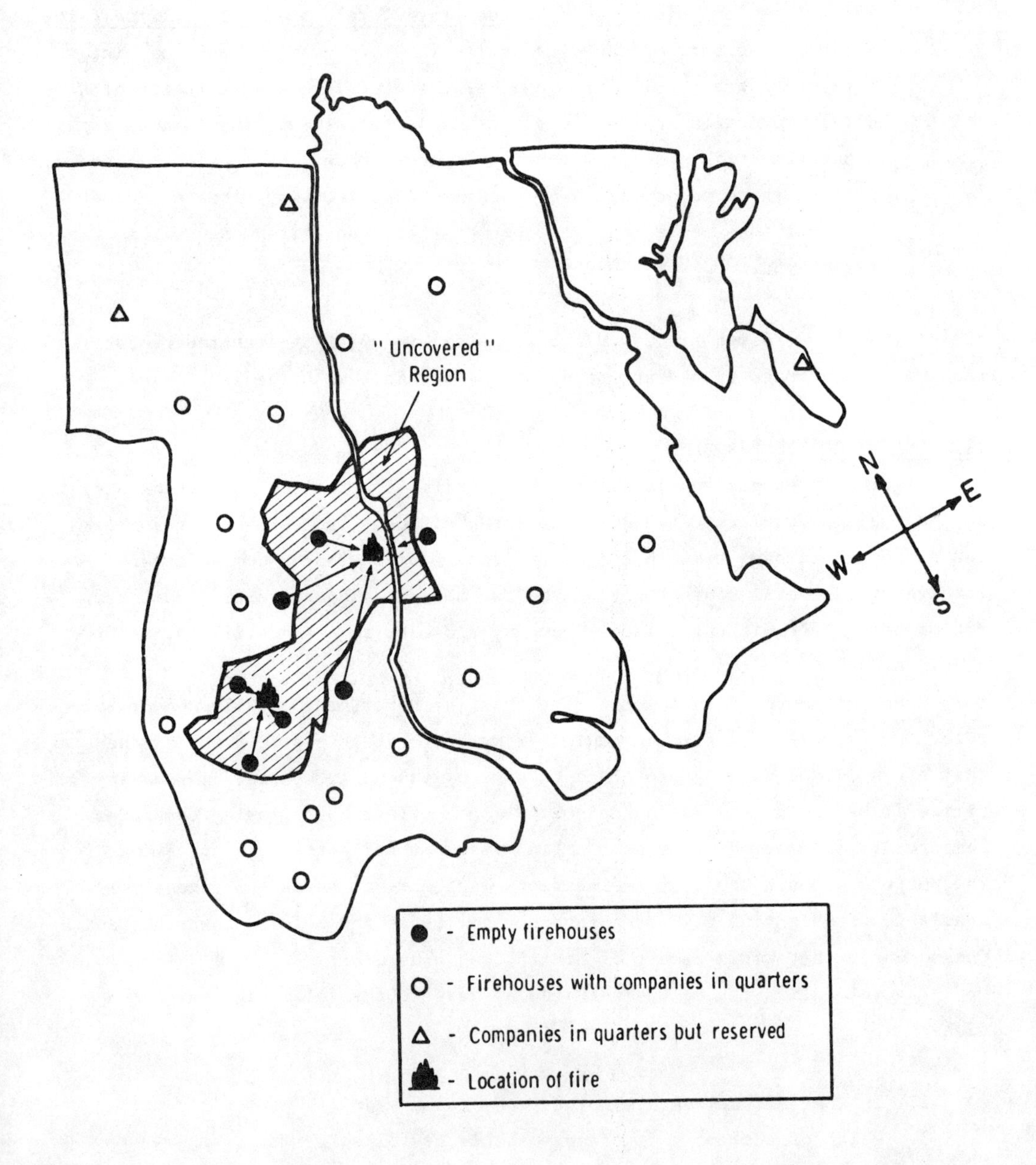

Fig. 1 - A sample relocation problem

possible to plan in advance for relocations with a reasonable expectation that the plan will be able to be carried out. Experienced fire officers imagine a hypothetical incident, say a three-alarm fire at a particular alarm box. Using their judgment, and assuming that all other fire companies not called to the third alarm will be available, they formulate a specific relocation plan. The plan consists of a list of temporary transfers of engine companies and ladder companies. Figure 2 is an example of an alarm assignment card used for dispatching and relocating fire companies in New York City. There is one of these cards for every alarm box in the city (over 14,000). The left side of the card shows the units (engine companies, ladder companies, and battalion chiefs) that are to be dispatched to alarms received from that box. The right hand side shows the companies that are to relocate when there is a serious fire in progress near that location. For example, the second line (corresponding to a "two-alarm" fire) shows that Engine 50 would be moved into the house of Engine 75, Ladder 49 would be moved into the house of Ladder 33, etc. When the alarm rate is low, alarm assignment cards work well. When the alarm rate is high, however, the plans often break down. The reason for the breakdown is simply that at high alarm rates several incidents (including small fires) may be in progress simultaneously, and the officers who created a relocation plan for one particular incident could not have anticipated this. To make a good and implementable relocation in this situation requires knowledge of the status of all the fire companies at the department's disposal and the nature of all incidents in progress *at the time action must be taken.* There are so many possible variations of the situation that can be encountered that there is no way to do this in advance.

Figure 3 shows two of the problems that can arise with pre-planned relocations.

1. A company that is supposed to relocate is already busy.
2. A company that is supposed to relocate is available, but moving it might create an even bigger gap in coverage.

We set out to design a new relocation procedure that could be used as part of an on-line real-time computer-assisted command and control system and that could be relied on to produce relocations that we, the fire department, and the public would agree were "good" (as opposed to "optimal") in all types of situations.

Our approach to developing a procedure was to view each of the questions raised above as a separate decision problem, with the solution to each problem being used as input to the next problem. We did this because the overall problem is a very large multiobjective problem that would be hard to implement

3311	CRESTON AVENUE and 192nd STREET							BRONX	
ENGINE CO'S	Marine Co.	Res. Co.	LADDER CO'S	D.C	B. C.	Special Apparatus	Covering Chiefs	COMPANIES TO CHANGE LOCATION 11/67	
								ENGINE	LADDER
48 75 79			33 37	7	19		B. C. 15		
81 88 42			46				D. C. 6	50-75 38-79	49-33 32-37
43 46 62 95		3	38		18			41-46 90-62 67-95	
92 45 68 93			27					83-92 94-45 80-68 59-93	19-27
82 71 60 69			36					96-82 35-71 53-60 40-69	34-36

Fig. 2 - Alarm assignment card for Bronx box 3311

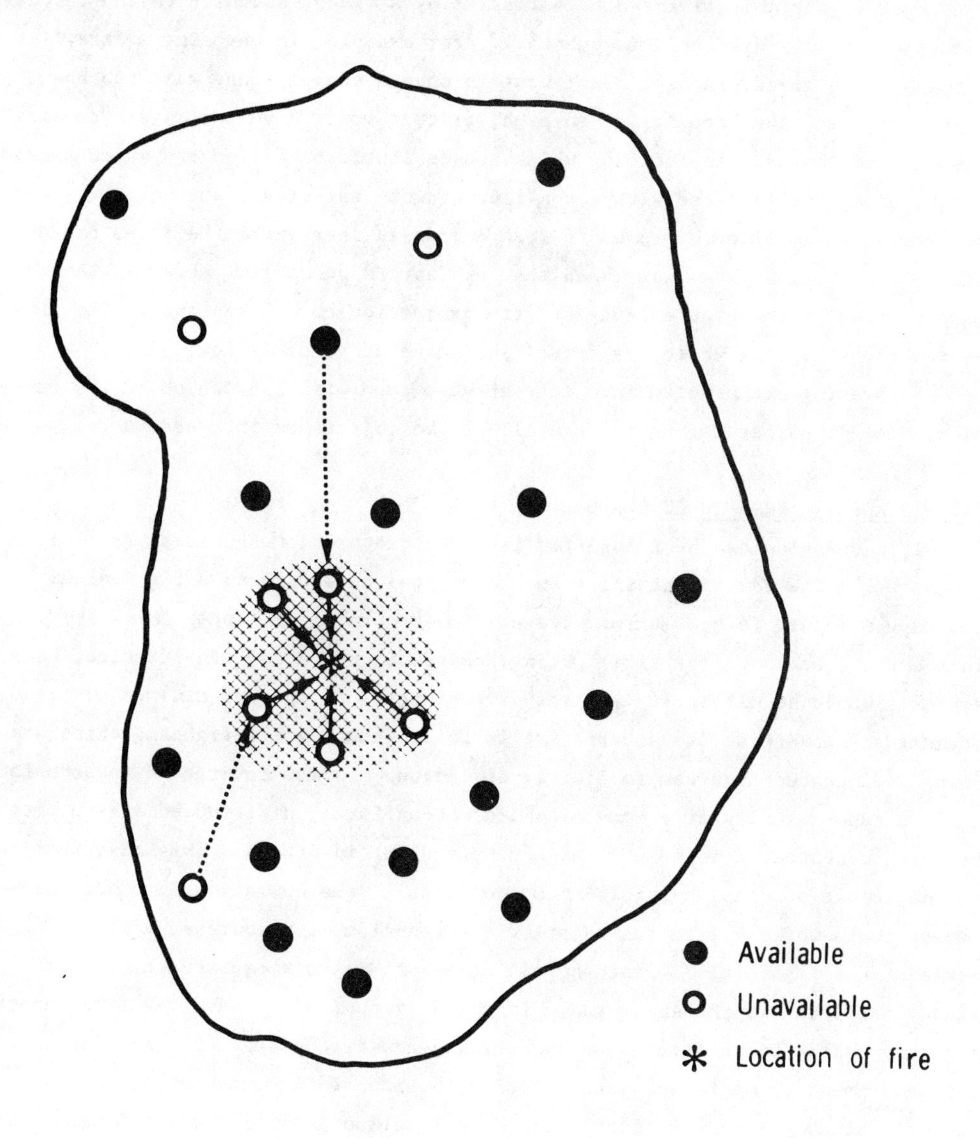

Fig. 3—Problems with the traditional relocation policy

within the computer time and space constraints of a real-time system (which would be performing other functions at the same time).

Most of the objectives to be satisfied by a "good" solution to the relocation problem were difficult to quantify. For example, in choosing a fire company to be moved, one would not want to move a company that was "too busy," that was protecting "too large" an area, or that would have to travel "too far" from its own house. In deciding which houses should be filled and which should remain empty one is faced with a conflict between efficiency and equity. The department would like to place its fire companies near where the fires are expected to occur, so they can reach them as fast as possible. However, they must provide an acceptable level of fire protection to all regions of the city --even those regions where the actual incidence of fires is low.

Separating the problem into four stages, each with its own objective function, made it easier for us to take all of the objectives into account.*

III. STAGE 1: WHEN TO RELOCATE

This question can be translated into the question, "when is an area underprotected?" One way of obtaining an operational answer is to set a minimum coverage standard (e.g., maximum travel time or travel distance) for every point in the city, based on the firefighting demands in the area. In practice, however, it would be difficult (and rather arbitrary) to specify minimum coverage standards. An attractive alternative is to let the way firefighting units are already allocated to areas implicitly define the minimum coverage standards for those areas. Usually fire companies are not uniformly distributed over a city but are concentrated in some areas and spread out in others. This distribution is the result of complex forces--some political, some operational, others historical. In working with its existing distribution of resources, a fire department has implicitly decided how it wishes to balance equity against efficiency, at least in the short run. In the long run, of course, the fire department may modify the distribution by building new firehouses. By assuming that the department is satisfied with the distribution of fire companies, we can define a minimum coverage standard for the relocation problem that will maintain approximately the same relative geographic distribution of fire companies as currently exists. This, in turn, will maintain approximately the same variation in travel times (or distances) as currently exists between the areas.

Therefore, the coverage criterion that we used required that for every alarm box in the city, at least one of the k closest engine houses and at least one of the p closest ladder houses contain an available company. k and p are parameters that are set by fire department policy. New York City has used

---

*Complete details of the algorithm are contained in [6].

$k = p = 2$. Relocations are recommended whenever this coverage criterion is violated.

The application of this criterion is simplified by noticing that many alarm boxes will generally have the same k closest engines or p closest ladders. We call the aggregate of all alarm boxes having the same k closest engines an *engine response neighborhood* (written "engine RN" for brevity). A *ladder response neighborhood* ("ladder RN") is defined as the set of alarm boxes having the same p closest ladders. The set of engines and ladder response neighborhoods each form nonoverlapping partitions of the city. They are defined separately since the coverage standard is to be applied separately to engines and ladders in order to keep a balance of each unit type in each region. The definition of minimum coverage can now be restated as: *there must be no engine response neighborhood with all of its k engines unavailable and no ladder response neighborhood with all of its p ladders unavailable.*

The use of response neighborhoods considerably reduces the calculations required to check on coverage. For example, in the Bronx there are over 2000 alarm boxes but with $p = 2$ fewer than 50 ladder RNs. Figure 4 shows the ladder RNs in the Bronx. Note that in regions where the ladder companies are close together the RNs are small, and where the companies are far apart they are larger.

We have delayed until now giving a precise definition of unit "availability" for purposes of minimum coverage. It makes no sense to relocate a unit into the house of a company responding to (but not yet working at) an alarm, returning from an alarm, or due back soon from a working fire. Therefore, we consider a company to be unavailable only if it is working at a fire expected to last for a "considerable" length of time (in practice, more than one hour).

IV. STAGE 2: WHICH HOUSES TO FILL

The primary objective of the relocation algorithm is to maintain minimum coverage as just defined. It makes sense to do this by moving as few companies as possible, since moving companies increases communication problems, places them in regions with which they may not be familiar, and takes them away from their home bases, food, and dry clothes. So we take as our criterion for the determination of empty houses to fill: have every response neighborhood covered, but move as few companies as possible. This translates into an integer program known as the set covering problem.

Suppose there are K uncovered RNs and L vacant houses whose busy companies cover or serve these RNs. For our decision variables let $x_j = 1$ if house j is to be filled and $x_j = 0$ otherwise. Then the problem is:

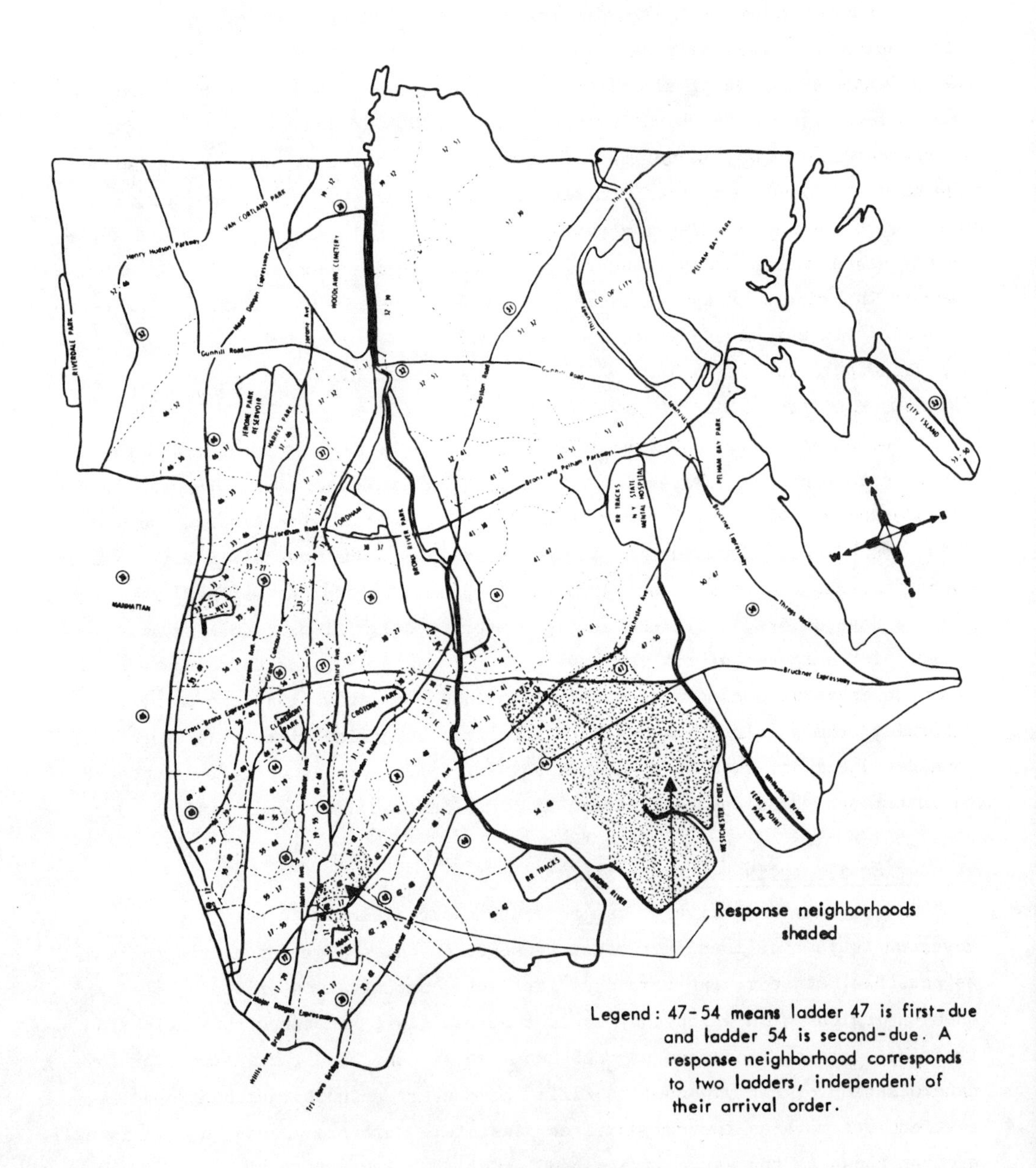

Fig. 4 — Ladder response neighborhoods in the Bronx

$$\text{minimize} \quad \sum_{j=1}^{L} x_j$$

$$\text{subject to} \quad \sum_{j=1}^{L} a_{ij}\, x_j \geq 1 \qquad i = 1, 2, \ldots, K$$

$$x_j = 0, 1 \qquad j = 1, 2, \ldots, L$$

$$\text{where } a_{ij} = \begin{cases} 1 & \text{if the jth house's busy company covers or serves the ith RN,} \\ 0 & \text{otherwise} \end{cases}$$

The matrix of the a_{ij} is known as the *incidence matrix* for the covering problem. Each row of the matrix corresponds to an uncovered RN. There will be p elements equal to 1 in each row when we consider ladders, and k equal to 1 in each row when we consider engines. The columns correspond to the houses of the unavailable companies. There will be a 1 in each row-column position that marks the correspondence between an unavailable company and a RN it is supposed to cover when it is available. The output of stage 2 is a set of M ($M \leq L$) vacant houses to be filled, which is the input to stage 3.

This problem can, of course, be solved exactly. However, for reasons of speed and computer storage requirements, we developed simple heuristic procedures for solution of this problem, and those encountered in the next two stages. In testing the algorithms, the results obtained using exact procedures were compared to those obtained using the heuristics. Fortunately, in the tests the optimal solution was always obtained using the heuristic methods.

The basic heuristic rule for selection of a house to fill is to select first the house associated with the largest number of uncovered RNs. After application of this rule, the covering problem is reduced by the elimination of the house just selected to be filled and all RNs that will be covered as a consequence of filling it. In the same way, another house is selected to be covered. This procedure continues until all RNs are covered. The rule may be applied several times using alternate starting points, and the best of the results chosen.

V. STAGE 3: WHICH AVAILABLE COMPANIES TO MOVE

Once the houses to be filled have been selected on the basis of the minimum coverage criterion (which is basically an equity criterion), there may be many available companies that could be moved into those houses. Of course, no company should be moved if, by moving it, the minimum coverage criterion is violated. We therefore only consider moves that do not violate the standard. In addition, we wish to apply the following secondary "efficiency" criteria:

1. Do not move a company "too long" a distance.
2. Do not move a company that is "too busy."
3. Do not move a company that is protecting "too big" an area.

The way to tradeoff among these criteria is not immediately clear. A function that measures travel time was found to take all these secondary factors into account. We end up with a mathematical programming problem whose objective is to minimize the total expected travel time to alarms that occur in the regions affected by the moves. As in stage 3, the mathematical programming problem is solved heuristically.

The following simplified scenario provides the background for the development of the objective function. Referring to Fig. 5, suppose Ladder 31 has just responded to a serious fire. Its house is now empty and we wish to evaluate possible relocations into it. The houses of Ladder Companies 37 and 38 are currently covered, and either one may be moved into Ladder 31's house. We want to evaluate which move is superior and if indeed any move should be made.

First, consider the data required to make the evaluation. Let R_{31} denote the region in which Ladder 31 would be the closest company to all alarm boxes if it were available in its house. We must take into account which of Ladder 31's neighbors are currently available when we decide what constitutes R_{31}. Let A_{31} and λ_{31} denote respectively the physical area and the alarm rate of this region. (Note that this region changes as the pattern of available and busy companies changes.) Let R_{37} denote the region in which Ladder 37 is *currently* the closest available company, and A_{37} and λ_{37} be, respectively, the area and alarm rate in R_{37}. Once again, in deciding what constitutes R_{37} we must take into account which of Ladder 37's neighbors are available. Similar definitions apply for R_{38}, A_{38}, and λ_{38}. We assume that the alarm arrivals are a Poisson process and that all the above parameters have been estimated. In addition, r_{ij}, the time required to relocate a company from location i to location j, is assumed to be known.

The expected travel time of the first-arriving company to an alarm in the regions served by these companies depends on which companies are available to respond to the alarm. If the closest company is not in quarters, the second closest company must play the role of the closest company, and consequently the travel time will be longer. The computational requirements for the exact calculation of expected travel times in such a dynamically changing environment are formidable. We therefore used an approximate method based on two models*:

*The theoretical and empirical justification for the use of these models is provided in [5] and [7].

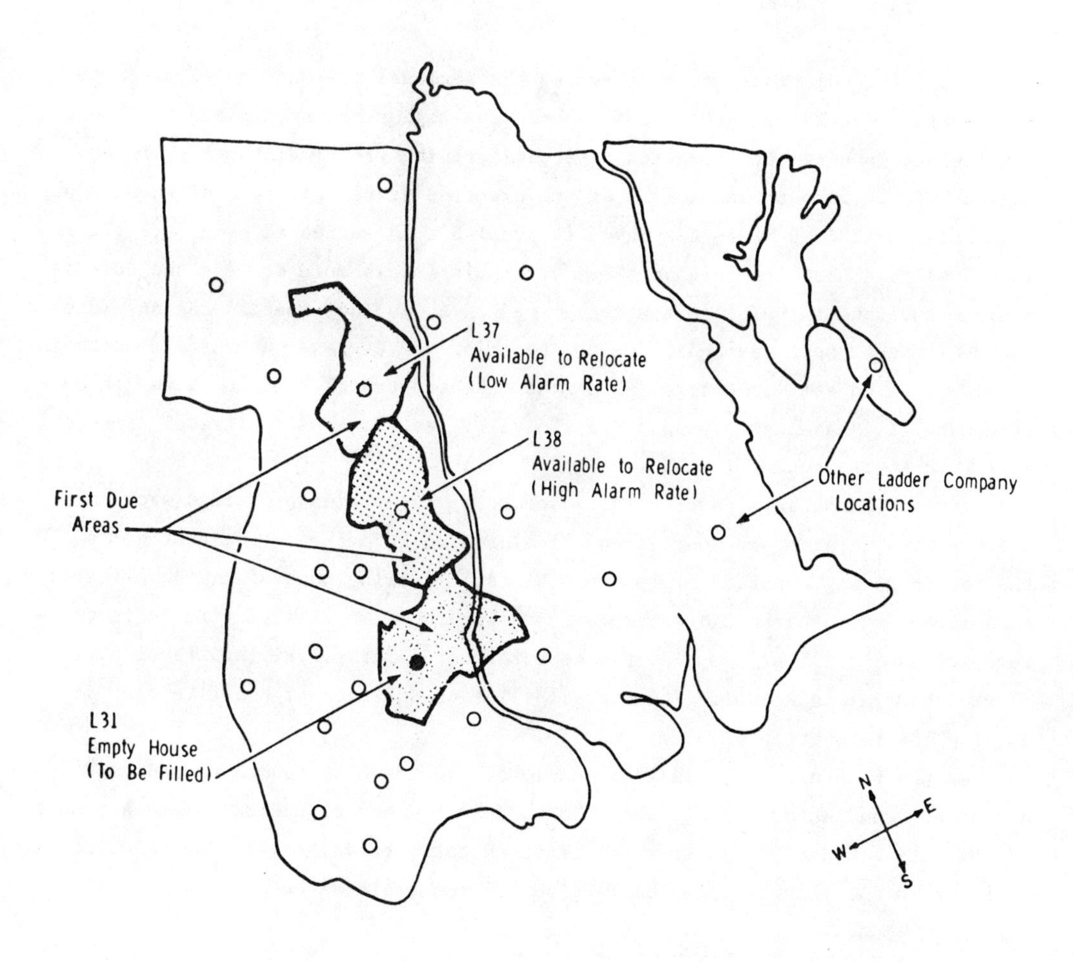

Fig. 5 - A simple relocation problem: Which of the two available companies should fill the empty house?

1. average travel distances in a region are directly proportional to the square-root of the area of the region and inversely proportional to the number of companies located in the region; and
2. average travel times in a region increase linearly with the average travel distance.

The travel-distance model estimates the regional average travel distance for the first-arriving unit as $c_1\sqrt{A/N}$ when the region has area A and has N units available. To estimate the average second-arriving travel distance c_1 is replaced by c_2. In our case N = 1, so the expected travel distance of the closest responding unit in R_i when Company i is available is estimated by $c_1\sqrt{A_i}$. If it is unavailable, but its neighbors are available (as is more or less the case if minimum coverage is being guaranteed), then $c_2\sqrt{A_i}$ is the expected response distance of the closest responding unit. Denoting the average response velocity in R_i by v_i, we have $\tau_1(i) = c_1\sqrt{A_i}/v_i$ as the expected travel time to alarms in R_i if Company i is available, and $\tau_2(i) = c_2\sqrt{A_i}/v_i$ as the expected travel time to alarms in R_i if Company i is busy.

If the alarms in R_i are arriving according to a Poisson process with an average of λ_i alarms per hour, then $\lambda_i T$ alarms would occur on the average in the region during a period of length T hours. Ignoring some of the complicated dynamic behavior that could occur in R_i during the time a large fire is in progress, we have $\tau_1\lambda_i T$ or $\tau_2\lambda_i T$ as the expected total first-arriving travel time to alarms occuring in R_i during the interval T--the duration of the fire that is causing the relocation problem.

We now return to the simple scenario of Fig. 5 where Ladder Companies 37 and 38 are candidates to relocate into Ladder 31's house, and calculate the cost of relocating Ladder 37 in terms of expected total travel time. The relevant information for this calculation is given in the table below.

i	$r_{i,31}$ (hours)	λ_i (alarms/hr.)	A_i (sq. miles)	$\tau_1(i)$ (mins.)	$\tau_2(i)$ (mins.)
37	0.2	0.2	0.9	1.7	2.8
38	0.1	1.2	1.3	2.1	3.5
31	--	1.7	0.9	1.8	2.9

Assume that the fire at which Ladder 31 is working will last one hour (i.e., T = 1 hour). In order to compare the cost of relocating Ladder Company 37 to the cost of relocating Ladder Company 38 we will consider alarms that occur in the interval $T' = [0, T + \max(r_{37,31}, r_{38,31})] = [0, 1.2]$. That is, T' has to be long enough to encompass all of the effects of any relocations in the system.

The cost of moving Ladder 37 to Ladder 31 ($c_{37,31}(T')$) is based on the assumption that Ladder 37 spends a time $r_{37,31}$ traveling to Ladder 31's house, stays at that house until Ladder 31 returns from the fire at time T, and then returns home. So R_{31} is covered by a second-closest company during the interval $[0,r_{37,31}]$ and by a closest company during the interval $[r_{37,31},T']$; R_{37} is covered by a second-closest company during the interval $[0,T']$; and R_{38} is covered by a closest company during the interval $[0,T']$. We are using an important property of the Poisson process, namely, that when two or more independent processes are observed simultaneously, the "joint" process (the results of both taken together) is also a Poisson process and has as its rate the sum of the rates of the individual processes. The components of the cost are:

For Ladder 37's area:	$\tau_2(37)\lambda_{37}T' = (2.8)(0.2)(1.2)$	$= 0.67$
For Ladder 31's area:	$\tau_2(31)\lambda_{31}r_{37,31} + \tau_1(31)\lambda_{31}(T' - r_{37,31})$	
	$= (2.9)(1.7)(0.2) + (1.8)(1.7)(1.0)$	$= 4.05$
For Ladder 38's area:	$\tau_1(38)\lambda_{38}T' = (2.1)(1.2)(1.2)$	$= 3.02$
Total cost (minutes)	$= c_{37,31}(T')$	$= 7.74$

If Ladder 38 relocated into Ladder 31's house, the expected total first-arriving ladder travel time over the interval $[0,T']$ would be calculated in a similar fashion. The result is that $c_{38,31}(T') = 9.14$ minutes, which is significantly higher than $c_{37,31}(T')$. The cost of making no relocation can also be calculated. In this case the cost would be 9.35 minutes. So, in this case, the best policy would be to relocate Ladder 37 into the house of Ladder 31.

In general, if we let $\alpha_k = \lambda_k\sqrt{A_k}/v_k$, we can show that the cost (in expected total travel time) of relocating available company i into the empty house of company j is given by

$$c_{ij}(T) = (c_2 - c_1)\{\alpha_i(T + r_{ij}) + \alpha_j r_{ij}\},$$

and the cost of making no relocation is just $(c_2 - c_1)\alpha_j T$.

Notice that each of the three secondary criteria—relocation travel distance (r_{ij}), the "busyness" of a company (λ_i), and the size of the region protected by a company (A_i)—are all explicitly included in the cost function. In addition, another element appears that perhaps had not been anticipated: the duration of the fire causing the relocation problem. According to the cost function, it is possible that a different relocation would be suggested for a short incident than for a long incident. In fact, using this function it is possible to determine what the predicted length of the incident must be before it becomes advantageous to relocate.

Figure 6 illustrates the typical relocation costs for the situation in which Ladder Company 31 is working at a fire and Ladder Companies 37 and 38 are available to relocate. The average total first-arriving ladder travel time is shown for fires that occur during the duration of the incident leading to the relocation for four alternatives:

1. No relocation (Ladder 31's house remains uncovered);
2. Move Ladder 38, which is closer to Ladder 31, but is a busy company;
3. Move Ladder 37, which is farther away from Ladder 31, but is less busy;
4. Relocate Ladder 37 into Ladder 38's house, and relocate Ladder 38 into Ladder 31--called a "successive moveup."

For any given value of T, the best relocation is the one for which the function is the smallest. Examination of the graph indicates that it certainly does not pay to make a relocation for an incident of predicted duration less than 15 minutes. If the fire will last longer than that, the best plan is to move Ladder 37 into Ladder 31's house. If the incident lasts more than a half hour, there is a clear advantage to this relocation. Note that if Ladder 37 could not be moved it would not be worthwhile to make any relocation for a fire lasting a half hour or less. Figure 6 also shows that the successive moveup of L37 to L38 to L31 is slightly worse than just relocating Ladder 37 to Ladder 31's house.

Of course, these observations depend on the characteristics of the particular problem we have been examining. Nevertheless, examination of cost functions for many situations suggests some generalizations. First, "successive moveups" have about the same travel-time cost as simple relocations. They are sometimes a little better, more often a little worse. But they also move twice as many companies, thereby increasing the inconvenience to the men--as well as increasing communication and control problems. For these reasons, and after consultation with fire department senior officers, successive moveups were eliminated from further consideration in New York City. Second, in most situations in the city there seemed to be a clear travel-time advantage to relocating only if the relocation were to last more than one hour. (In the example, the advantage showed at about a half hour, but this is due to the fact that Ladder 31 was at that time the busiest ladder company in the city.) In general, an incident that will last an hour or more is identifiable by a chief when he first arrives at the fire. So, rather than requiring the exact duration of all incidents (the value of T in the cost function), it was suggested that relocations be made only for fires expected to last more than one hour.

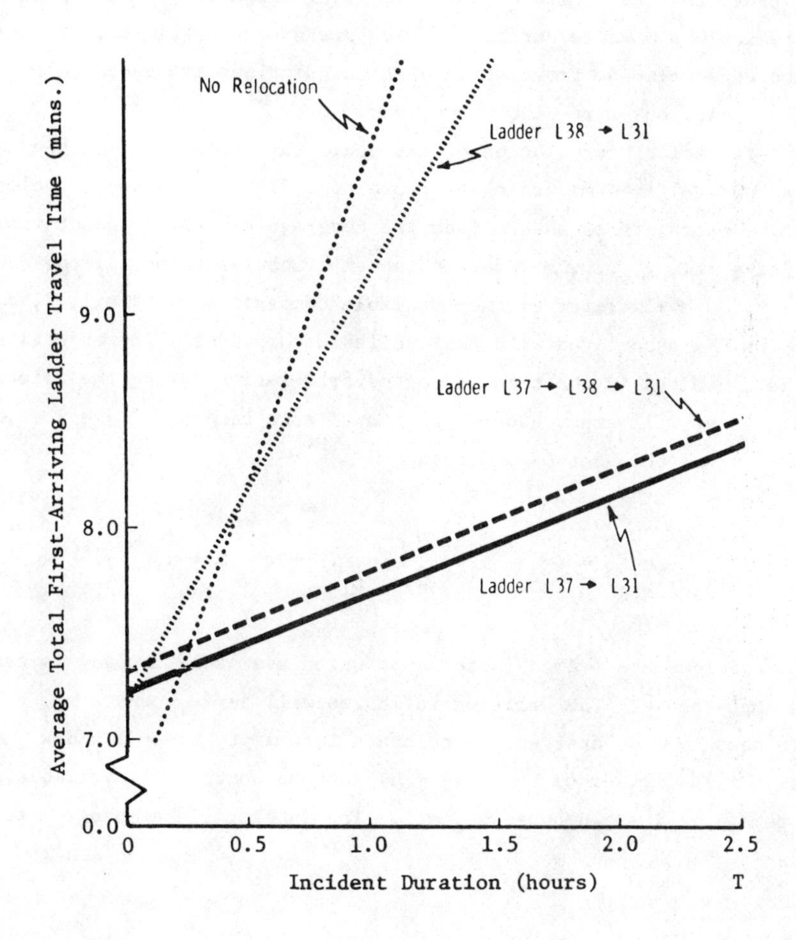

Fig. 6--A comparison of relocations into the house of Ladder 31

The actual duration of such "serious fires" does not change the *relative* travel-time cost rankings very much. That is to say, if moving Ladder 37 to the Ladder 31 house looks better than moving Ladder 38 there for a one-hour fire, it also looks better for a two-hour fire. This is important since, if it were not true, the identities of the relocating companies would depend on accurate predictions of fire duration (which are not generally easy to make). In the method we developed, travel-time cost calculations are based on the average duration of a serious fire--about one hour.

The mathematical formulation of the stage three problem (which available companies to move) is similar in structure to a "transportation" problem with additional constraints to assure that the coverage criteria are not violated.

We let $j = 1, 2, \ldots, M$ refer to the empty houses to be filled, $j = M + 1, \ldots, M + N$ refer to the available companies, and $k = 1, 2, \ldots, L$ refer to the RNs associated with the available companies. The objective function to be minimized is the total expected travel-time during the relocation incident. If we fill empty house j with available company i_j $(j = 1, 2, \ldots, M)$, the total cost of this set of moves is

$$\sum_{j=1}^{M} c_{i_j j} = (c_2 - c_1) \sum_{j=1}^{M} \{\alpha_{i_j}(T + r_{i_j j}) + \alpha_j r_{i_j j}\}$$

The decisions are framed in terms of which available company is assigned to which empty house. The decision variables will be x_{ij}, where $x_{ij} = 1$ if available company i is assigned to relocate into empty house j and $x_{ij} = 0$ otherwise. A "dummy" empty house $(j = 0)$ is used so that if available company i is to remain in its own house, $x_{i0} = 1$. The integer linear program to be solved is:

$$\text{min.} \quad \sum_{j=1}^{M} \sum_{i=M+1}^{M+N} c_{ij} x_{ij}$$

$$\text{s.t.} \quad \sum_{i=M+1}^{M+N} x_{ij} = 1 \qquad j = 1, 2, \ldots, M$$

$$\sum_{j=0}^{M} x_{ij} = 1 \qquad i = M+1, M+2, \ldots, M+N$$

$$\sum_{i=M+1}^{M+N} a_{ik} x_{i0} + \sum_{i=M+1}^{M+N} \sum_{j=1}^{M} a_{jk} x_{ij} \geq 1 \qquad k = 1, 2, \ldots, L$$

$$x_{ij} = 0, 1 \qquad \text{for all } i, j.$$

The objective function and the first two sets of constraints have the structure of a transportation problem. The first set of constraints requires that all M of the empty houses are filled. The second set of constraints guarantees that all available companies are assigned somewhere. As in stage 2, the coefficients a_{ij} form the incidence matrix between firehouses and response neighborhoods. $a_{ik} = 1$ if available company i serves response neighborhood k and is zero otherwise. The last set of constraints requires that none of the RNs that were being covered by available companies be uncovered.

Rather than solve such a large integer programming problem exactly, we use a heuristic algorithm. The heuristic rule that we use for determining the available company to move into a given house is to try to fill each empty house with the available company having the lowest relocation cost associated with the move. The relocation costs are the c_{ij} described above. To help select the relocatees (companies to be relocated) a ranked list of candidate relocatees can be created for each house to be filled. The companies on the list are the available ones, ordered by their c_{ij} values.

Each move must be checked against the coverage criterion to assure that no RNs become uncovered. A company on any relocatee list may be relocated without violating minimum coverage. But, if the selections are made independently for each vacant house to be filled, the resulting set of moves might have the same company moving into more than one house, or might leave one or more RNs uncovered by moving neighboring companies.

A feasible relocation is generated by successive applications of the heuristic and the feasibility test. The procedure begins with one house to be filled and progresses in sequence through the others one by one. If the lowest cost move for each house produces a feasible relocation, that relocation is optimal and no further computations are necessary. Otherwise, since the algorithm is fast, several feasible relocations are produced by changing the order in which houses being filled are considered, and by changing the heuristic for the first house being considered to "choose the available company with the qth lowest relocation cost." The least-cost relocation generated after all trials is used as the stage 3 solution.

VI. STAGE 4: SPECIFIC RELOCATION ASSIGNMENTS

The output of stage 3 is a specific set of assignments or "moves" of available companies to the empty houses being filled. However, the assignments sometimes make the relocating companies travel further than another possible assignment of the same set of companies to the same empty houses. In some rare instances the assignment can even make relocating companies travel on paths that cross. This is because travel distance is only one of the components of

c_{ij}. While the relocating travel distance matters, it is actually distance times alarm rate times the square-root of area that is being considered, and so the algorithm can sacrifice relocation travel distance for gains in expected travel times to alarms.

Yet for several reasons fire departments are concerned with the distance that relocating companies must move. One reason is that shorter relocation distances mean less of a burden on relocating companies and larger availability times. Another is that keeping the relocation distance down tends to keep companies in areas where they are familiar with street patterns as well as with particular firefighting problems. To solve this problem we view stage 3 as a device for selecting the companies to relocate and ignore the specific moves it suggests. We then determine the specific assignments that minimize total travel distance. Of course this would not make sense if the overall relocation cost were much higher. But in most cases the resulting "reassignment" increases the relocation cost very little, and can significantly reduce the total distance traveled.

We now let the index j = 1, 2, ..., M refer to the M *empty houses selected by stage 2* and the index i = 1, 2, ..., M refer to the M *available companies selected by stage 3*. Again, r_{ij} denotes the time required for a unit to relocate from "full" house i to "empty" house j, and the decision variable $x_{ij} = 1$ if available company i is assigned to empty house j and is zero otherwise.

Mathematically, stage 4 involves solving a traditional assignment problem:

Find $\{x_{ij}\}$ to

$$\min. \sum_{j=1}^{M} \sum_{i=1}^{M} r_{ij}\, x_{ij}$$

$$\text{s.t.} \quad \sum_{j=1}^{M} x_{ij} = 1 \qquad i = 1, 2, \ldots, M$$

$$\sum_{i=1}^{M} x_{ij} = 1 \qquad j = 1, 2, \ldots, M$$

$$x_{ij} = 0, 1.$$

For small problems, say up to M = 5, solutions can be obtained quickly by complete enumeration of all M! permutations (5! = 120). In fact it would be unusual to have to make even five relocations at one time. Usually a fire progresses through a number of stages (first alarm, second alarm, third alarm, etc.)

At each stage the number of relocations needed would be small. For example, the hypothetical case of two serious fires that break out simultaneously in the South Bronx, which was presented in Section II, led to the need for only four relocations. Figure 7 shows both the least cost set of relocations resulting from stage 3 of the algorithm (the dotted lines), and the least travel distance solution produced by stage 4 (the solid lines). The solution produced by stage 4 results in a reduction of 22 percent in travel distance with only a 9 percent increase in the cost function.

VI. TESTING AND IMPLEMENTATION

Before the New York City Fire Department would consider using the relocation algorithm in its Management Information and Control System, it insisted on extensive testing. We subjected the algorithm to a number of tests, first with problems that were designed to present difficult or interesting situations, second in a simulation model in which over 3600 alarms were generated at random according to historical patterns, and third, to provide a strenuous realistic test, in specifying relocations for one of the worst evenings ever experienced in the Bronx. In the last test, the sequence of incidents was reconstructed and a simulation was run to determine what would have occurred if the relocation algorithm had been operating during one of the most trying periods in recent departmental history. Finally, the algorithm was run in parallel with the existing manual system in one of the fire department's communications offices.

The problem whose heuristic solution is shown in Fig. 7 was also solved using an exact integer programming computer code. The result obtained was identical to the minimum cost solution found by the heuristic algorithm. However, the heuristic required only one-quarter of the CPU time and only one-half the amount of computer core storage.

When the algorithm was tested using a simulation model that recreated the actual situation in the Bronx on July 4, 1969, it was found that, if it had been operating at the time, it would have avoided almost all of the problems that actually occurred that evening. For example, in the actual situation at one point in the evening only 79 percent of the alarm boxes in the Bronx had at least one of their two closest ladder companies available in quarters. Using the algorithm over 90 percent of the boxes did. Of course, the algorithm never leaves a response neighborhood without "minimum coverage." However, on that night, a total of 16 RNs were actually left uncovered for periods ranging from 30 minutes to 1.6 hours. In addition, the algorithm generated its relocations gradually and continually over time, while relocations made by the dispatchers were generally made in spurts. For example, at one point both methods had produced 23 relocations but the algorithm had called for relocations to be made at

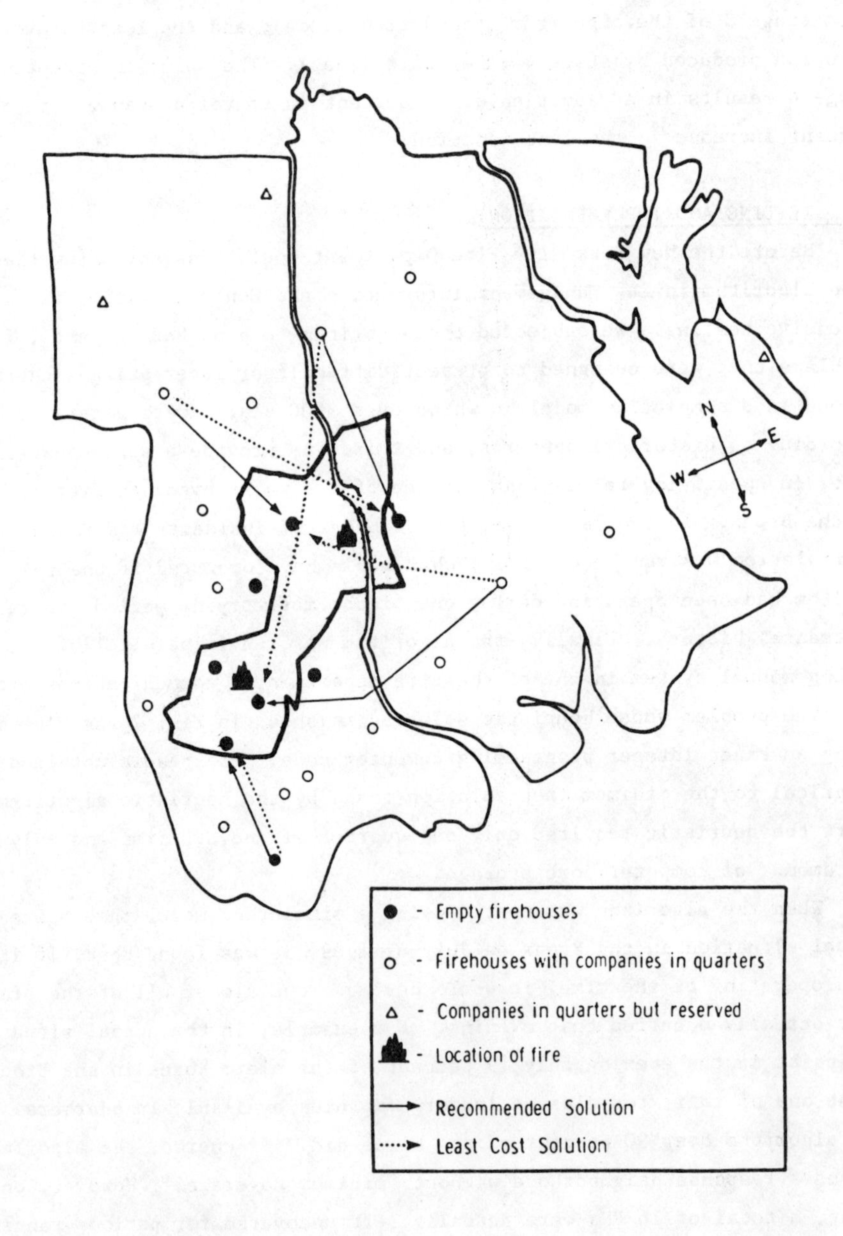

Fig. 7 - A comparison of the solutions from stages 3 and 4

16 separate times, while the dispatchers had made their relocations at only five different times. (Ten of their relocations were made at one time.)

The relocation algorithm was implemented as part of the New York City Fire Department's real-time computer-based Management Information and Control System in the middle of 1977. In addition to recommending relocations, the MICS recommends dispatches, maintains the status of fire companies and alarms in progress, and updates statistical records. The system, which was first implemented in the Brooklyn communications office, has now been expanded citywide [8].

REFERENCES

1. Carter, Grace, Jan Chaiken, and Edward Ignall, "Response Areas for Two Emergency Units," *Operations Research*, Vol. 20, No. 3, 571-594 (1972).

2. Chaiken, Jan, "Transfer of Emergency Service Deployment Models to Operating Agencies," *Management Science*, Vol. 24, No. 7, 719-731 (1978).

3. The Rand Fire Project (Warren Walker, Jan Chaiken, and Edward Ignall, editors), *Fire Department Deployment Analysis*, Elsevier North-Holland, New York (1979).

4. Rider, Kenneth, "A Parametric Allocation Model for the Allocation of Fire Companies in New York City, *Management Science*, Vol. 23, No. 2, 146-158 (1976).

5. Kolesar, Peter, and Edward Blum, "Square Root Laws for Fire Engine Response Distances," *Management Science*, Vol. 19, No. 12, 1368-1378 (1973).

6. Kolesar, Peter, and Warren Walker, "An Algorithm for the Dynamic Relocation of Fire Companies," *Operations Research*, Vol. 22, No. 2, 249-274 (1974).

7. Kolesar, Peter, Warren Walker, and Jack Hausner, "Determining the Relation Between Fire Engine Travel Times and Travel Distances in New York City," *Operations Research*, Vol. 23, No. 4, 614-627 (1975).

8. Mohan, John J., "Starfire, F.D.N.Y.," *With New York Firemen*, Vol. 41, No. 1, 12-13 (1980).

9. Swersey, Arthur, "A Markovian Decision Model for Deciding How Many Units to Dispatch," in *Models for Reducing Fire Engine Response Times*, Ph.D. Thesis, Columbia University, New York (1972).

10. Walker, Warren, *Firehouse Site Evaluation Model: Executive Summary*, Report R-1618/1-HUD, The Rand Corporation, Santa Monica (1975).

BCDEFGHIJ-AMS-89876543